Imaging Industrial Flows

The authors are grateful to the following for granting permission to reproduce figures included in this book:

The authors of all figures not originated by ourselves
Academic Press
American Chemical Society
American Institute of Chemical Engineers
BHR Group Limited
Butterworth-Heinemann Ltd
Cambridge University Press
Elsevier Science Ltd, Pergamon Imprint
Institute of Chemical Engineers
IEEE
Macmillan Press
McGraw Hill
UMIST, Department of Electrical Engineering and Electronics
Shell Research Limited
UKAEA

The authors and Institute of Physics Publishing Ltd have attempted to trace the copyright holder of all the figures reproduced in this publication and apologise to copyright holders if permission to publish in this book has not been obtained.

Imaging Industrial Flows
Applications of Electrical Process Tomography

A Pląskowski

Micromath Ltd, Poland (Instrumentation)

M S Beck

Department of Electrical Engineering and Electronics
University of Manchester Institute of Science and Technology

R Thorn

Department of Fluid Engineering and Instrumentation
Cranfield University

T Dyakowski

Chemical Engineering Department
University of Manchester Institute of Science and Technology

Institute of Physics Publishing
Bristol and Philadelphia

British Library Cataloguing-in-Publication Data

A catalogue record for this book is available from the British Library.

ISBN 0 7503 0296 8

Library of Congress Cataloging-in-Publication Data

Imaging industrial flows : applications of electrical process
    tomography / A. Pląskowski ... [et al.].
        p.   cm.
    Includes bibliographical references and index.
    ISBN 0-7503-0296-8.
    1. Tomography--Industrial applications. 2. Flow visualization--Industrial applications.   I. Pląskowski, A. (Andrzej)
    TA417.25.I39 1995
    681'.2--dc20   95-3373
                                                                    CIP

Published by Institute of Physics Publishing, wholly owned by The Institute of Physics, London

Institute of Physics Publishing, Techno House, Redcliffe Way, Bristol BS1 6NX, UK

US Editorial Office: Institute of Physics Publishing, The Public Ledger Building, Suite 1035, Independence Square, Philadelphia, PA 19106, USA

Printed in the UK by J W Arrowsmith, Bristol

To our wives for their love and support

Contents

Preface   xi

List of principal symbols   xvii

1   Why is flow imaging needed?   1
    1.1   The problem of flow measurement   1
    1.2   The importance of multi-component flow measurement   2
    1.3   Terminology   5
    1.4   Two-component flow measurement: the basic problem   5
    1.5   Conventional methods of two-component flow measurement   7
    1.6   Component velocity measurement   9
    1.7   Component concentration measurement   18
    1.8   Direct mass flow measurement   22
    1.9   The flow imaging method of two-component flow
          measurement   24
    1.10 The basic subsystems for flow imaging   28
    1.11 The need for a systematic approach to design   29
    1.12 Summary   29

2   Two-phase fluid dynamics   31
    2.1   Introduction   31
    2.2   Categories of flow images   31
    2.3   Overview of two-phase phenomena   34
    2.4   Macroscale flow phenomena   36
    2.5   Method for predicting flow regime transition for gas/liquid
          systems   42
    2.6   Microscale structure   54
    2.7   Mathematical modelling of two-phase flow   61
    2.8   Summary   74

3   Sensing techniques   76
    3.1   Introduction   76
    3.2   Classification of sensors   80

3.3	Measurement for flow imaging	82
3.4	Capacitance sensors—a major case study	83
3.5	Electrical impedance tomography for flow imaging of conducting fluids—an outline	105
3.6	Ultrasonic sensors—an outline	111
3.7	Summary	120

4   Image reconstruction   122

4.1	Introduction	122
4.2	Basic principles of image reconstruction	122
4.3	Two-component flow image reconstruction using a two-projection knowledge-based reconstruction algorithm	126
4.4	Reconstruction algorithm—capacitive impedance sensing technique	131
4.5	Reconstruction algorithms–resistive impedance sensing technique	135
4.6	Reconstruction algorithms—ultrasonic flow imaging	141
4.7	Summary	147

5   Image display and interpretation   148

5.1	Introduction	148
5.2	The need for image processing	148
5.3	Image presentation	150
5.4	The grey level histogram	151
5.5	Enhancement of an image using filtering techniques	152
5.6	The thresholding technique	158
5.7	Shrinking and expansion techniques	160
5.8	Measurements on objects in the image	161
5.9	Detection of all objects within the image frame	167
5.10	Blurring of images caused by backprojection	170
5.11	Summary	170

6   Applications   171

6.1	Introduction	171
6.2	Electrical capacitance tomography—performance parameters	172
6.3	Electrical capacitance tomography for imaging gas/solids flows in pneumatic conveyers	180
6.4	Electrical capacitance tomography for imaging gas flow in fluidized beds	183
6.5	Flow regime identification using electrical capacitance tomography sensing systems	184
6.6	Electrical resistance tomography for measuring transient concentration profiles	187

6.7   Summary   189

7   The future   **190**
7.1   The need for action   190
7.2   The new trends   192

References   **199**

Index   **207**

Preface

The tomographic imaging of objects offers a unique opportunity to unravel the complexities of structure without the need to invade the object. The development of tomographic instrumentation, started in the 1950s, has led to the widespread availability of body scanners, so much a part of modern medicine.

In the 1990s industry is under pressure to utilize resources more efficiently, and to satisfy demand and legislation for product quality and reduced environmental emissions. Hence there is an increasing need to know more about the exact way the internal flows in process equipment are behaving. Often this must be done non-invasively by tomographic instrumentation because conventional measuring instruments may either be unsuitable for exposure to the harsh internal conditions of the process, or by their presence upset the operation of the process. There is now a widespread appreciation of the need for the direct analysis of the internal characteristics of process equipment; the measuring instruments for such applications must use robust, non-invasive sensors which can operate in the proximity of aggressive and fast moving fluids and multiphase mixtures. In seeking an answer to this need the subject of flow imaging and the closely related subject of process tomography is being developed. This involves using tomographic imaging methods to manipulate the data from remote sensors in order to obtain precise quantitative information from inaccessible locations. Flow imaging will improve the operation and design of processes handling multi-component mixtures by enabling boundaries between different components in a process to be imaged in real time using non-intrusive sensors. Information on the flow regime, vector velocity, and component concentrations in process vessels and pipe lines will be determined from the images.

The opportunity to quantify the location and movement of various components in a pipeline offers the hope of designing non-intrusive multicomponent flowmeters, which are so sought after in many industries; especially in the oil industry for monitoring the output of subsea oil wells, which in future will have to use marginal resources of uncertain quality.

The basic idea is to install a number of sensors around the pipe or vessel to be imaged. The sensor output signals depend on the position of the component boundaries within their sensing zones. A computer is used to reconstruct a tomographic image of the cross section being observed by the sensors. This will provide, for instance, identification of the distribution of mixing zones in stirred reactors, interface measurement in complex separation processes and measurements of two-phase flow boundaries in pipes with applications to multi-component flow measurement.

Flow imaging essentially evolved during the mid-1980s. A number of applications of tomographic imaging of process equipment were described in the 1970s but generally these involved using ionizing radiation from X-ray or isotope sources and these were not satisfactory for the majority of applications on a routine basis, because of the high cost and safety constraints. Most of the radiation-based methods used long exposure times which meant that dynamic measurements of the real-time behaviour of flow in pipelines and process systems were not feasible.

In the mid-1980s work started that led to the present generation of flow imaging systems. At the University of Manchester Institute of Science and Technology in England (UMIST) we began a project on electrical capacitance tomography for imaging multicomponent flows from oil wells and in pneumatic conveyors. About the same time a group at the Morgantown Energy Technology Center in the USA were designing a capacitance tomography system for measuring the void distribution in gas fluidized beds. The capacitance sensors used for both the above systems were only suitable for use in an electrically non-conducting situation.

Also in the mid-1980s medical scientists started to realize the potential of electrical impedance tomography (measuring electrical resistance) as a safe, low-cost method for imaging the human body. There was rapid progress in several centres with the Sheffield University Royal Hallamshire Hospital in the UK and Rensselaer Polytechnic Institute in the USA taking major roles. The success of this early work encouraged the setting up in 1988 of a European Community (EC) co-ordinated activity on electrical impedance tomography for medical applications.

In 1988 work started at UMIST on the development of electrical impedance tomography (EIT) for imaging processes containing electrically conducting fluids. Previous work on medical EIT facilitated progress because some of the instrumentation problems were common to medical and to process EIT. However, major differences are apparent, principally medical EIT aims to measure the location of objects in space, whereas in process equipment there is a need to measure both the location and velocity of movement.

Undoubtedly, the 1980s development of low cost array processors has done much to solve the previous problem of high cost and slowness in tomographic image reconstruction using Von Neumann computer architectures and this has led to flow imaging becoming a cost-effective technique.

This book is mainly based on work at UMIST, where two new laboratories were set up in 1990, one in the Electrical Engineering Department concerned with research in instrumentation for flow imaging and the other in the Chemical Engineering Department concerned with industrial applications of the subject.

This prompted other UK-based research groups who began to meet on a regular basis in 1990. From these beginnings a European Concerted Action in Process Tomography was established. The first meeting was held in Manchester in 1992 and was attended by academic and industrial representatives from most European countries.

The need for tomographic instrumentation to explore and measure the flows in pipelines and other industrial equipment is now widely appreciated. In this book we present the basic principles of the sensors and image reconstruction techniques involved, together with application to case studies.

The reader is assumed to have a basic understanding of instrumentation, electronics and computing systems. Our overall aim is to present the current state-of-the-art in a concise form, but to include sufficient supporting information and theoretical description to enable the reader who is a student of the subject or an engineer newly commencing a research and development activity in process imaging to understand the subject thoroughly without reference to other books. The engineer with appropriate background knowledge who wishes to explore the potential of process imaging to solve industrial problems will probably omit Chapters 1 and 2 and concentrate on Chapters 3, 4, 5 and 6. The research engineer embarking on a more ambitious development programme may find Chapter 2 helpful for understanding the fluid dynamic behaviour of process systems in response to the operating conditions of the equipment.

Chapter 1 introduces the subject in terms of the industrial and economic need for flow measurement in general and for multi-component flow measurement in particular. The range of instrumental methods for measuring multi-component flows are discussed and the need for flow imaging is highlighted. The chapter concludes by summarizing the basic theory for multi-component flow measurement using imaging methods and showing how the flow imaging system can be broken down into three subsystems which are described later in the book.

Chapter 2 outlines the essential features of two-phase fluid dynamics which are needed for a complete understanding of the dynamic interactions between materials in a flow process system. Flow phenomena are described for a number of processes, including pipeline transport of mixtures, flow in fluidized beds and flow behaviour in shell and tube heat exchangers. This should help the reader to understand the way in which tomographic instrumentation can be used in the several related areas concerned with industries such as chemical engineering, oil technology and flow measurement instrumentation.

A systematic approach to the design of tomographic imaging systems is paramount to their successful applications, and in Chapter 3 we describe the development of the front-end sensing systems, with particular emphasis on the electrical sensing methods which are proving so successful because of their robustness, low cost and fast dynamic response. After discussing the relative merits of radiation-based and electrical sensing systems, we present a major case study of capacitance sensor design followed by outline studies of electrical resistive and ultrasonic sensing techniques for tomography.

In Chapter 4 we discuss the design of image reconstruction software. The relevant theory is outlined briefly, but we concentrate mainly on the description of backprojection reconstruction techniques which have been effective for the real-time imaging of industrial flows. The backprojection reconstruction techniques so far used do not exploit the potential for improved image quality that may be possible by using iterative image reconstruction techniques which, although being highly demanding on computer capacity, enable the sensor field distortion caused by high-contrast objects to be compensated for. Although iterative algorithms are mainly a subject for future development, their potential for providing a better image justifies their introduction in Chapter 4.

The way in which the image is displayed and interpreted to provide quantitative information on process behaviour is of importance in tomographic instrumentation. This is the subject of Chapter 5. In this Chapter we introduce ways in which the quality of images can be improved by image enhancement techniques and the methods which are available for obtaining quantitative data from images.

In Chapter 6 we describe some results which have been obtained from case studies using tomographic imaging systems for typical process applications.

Chapter 7 attempts to predict 'the future' of our subject. Not 'crystal ball gazing' on the work of a few individuals, but describing how the trend of a major international activity on Process Tomography is developing in the 1990s.

We wish to thank many people for their co-operation and help which has enabled us to write this book. Perhaps the greatest collective help was provided by the students and research staff in the Process Tomography Unit, previously in the Department of Instrumentation and Analytical Science and latterly in the Department of Electrical and Electronic Engineering at UMIST, and by academic and technical staff in these departments.

Certainly the most outstanding contribution to the preparation of this book has been made by our friend and colleague Dr A L Stott. He has read the draft text, and applied his penetrating technical approach to theoretical and practical matters. He has provided extensive descriptions to clarify difficult or obscure features. Our heartfelt thanks to you, Les, for all this painstaking work. Special thanks are due to Dr S M Huang, Dr C G Xie (now both with Schlumberger) and Dr F J Dickin for their technical contributions. We also thank Dr C G Xie for his detailed check of the final manuscript for flaws and omissions. Computer aided design methods, pioneered by Dr F Abdullah and Dr S Khan at City University, London, have enabled our imaging electrodes to be 'right first time'; we thank them for their painstaking work.

The Chemical Engineering Department at UMIST have pioneered the application of tomographic instrumentation systems and we especially thank Dr R A Williams for his untiring efforts in this connection. The practical applications have involved help from several industrial organizations; we would particularly like to thank Dr A Hunt and Dr C P Lenn of Schlumberger Cambridge Research and Mr B Edwards of Unilever for their painstaking and dedicated project management, and Mr R John of Tealgate for excellent engineering design.

Inter-university collaboration has been part of several projects. From the several groups it has been our privilege to work with, we would particularly single out the group at Leeds University led by Dr B S Hoyle, and that at Bergen University, Norway led by Professor E A Hammer, as representing the highest level of technical competence and co-operation.

The work reported in this book results from the resources made available by several organisations including the SERC, the British Council, the European Community, du Pont, ICI, British Nuclear Fuels, Schlumberger Cambridge Research, Shell and Unilever. Our secretary, Dorothy Denton, has handled details of the transition from untyped manuscript to final draft with skill and enthusiasm, and we gratefully thank her for this help. We thank Mr Rod Holt of the UMIST Audio-Visual Production Unit for assistance with preparing the diagrams.

Andrzej B Pląskowski, Warsaw
Maurice S Beck, Manchester
Richard Thorn, Cranfield
Tomasz Dyakowski, Manchester
1994

List of principal symbols

A	matrix of sensitivity factors
	gain
	area
	signal amplitude
c	the velocity of transmitted energy
C	electrical capacitance
d	gas bubble diameter
	distance
D	pipe line diameter
Eö	Eötvos number
F	Froude number
$f(\ )$	function
f	frequency
	friction factor
G	image grey level
$H(\ )$	transfer function
I	radiation intensity
	electrical current
	image plane
J	electric current density
k	viscosity ratio
	proportional coefficient
L	sensor spacing along axis of flow
M	number of independent measurements
	fluid group property defined by equation (2.25)
	mass flow rate
N	number of pixels
	number of electrodes/sensors
P	number of projections
Q	volumetric flow rate
	charge
R	electrical resistance
	Radon transform
	radius

Re   Reynolds number
$R_{(x,y)}(\ )$ cross correlation function of x and y
s the Laplace variable
              sheltering coefficient
t time
T time interval, time constant, period
U velocity
v velocity
V voltage
              vector of views
$w(\ )$ weighting function
(x, y) Cartesian co-ordinates
x vector representing the coefficients of the object space
y vector of data measured by sensors
α component ratio/hold-up/void fraction
              angle
β angle of inclination
γ bubble distortion coefficient
              average value of normalized capacitance measurements
ϵ permittivity/dielectric constant
ζ average value of grey level of normalized original reconstructed image
η threshold level
Θ angle
              phase difference
μ total attenuation coefficient
              viscosity
              permeability
μ_a absorption attenuation coefficient
μ_s scattering attenuation coefficient
ρ density
              resistivity
σ surface tension
              conductivity
τ time delay, transit time, time interval
ϕ angle
Φ electrostatic potential
ω angular frequency
$\langle\ \rangle$ an average

1

Why is flow imaging needed?

1.1 THE PROBLEM OF FLOW MEASUREMENT

It is over 20 years since advances in technology allowed man to land on the moon, and yet for many applications the apparently simple task of accurately measuring how much fluid is flowing through a pipe has still to be satisfactorily solved. Why should this be so? Maybe the problem is just not an important one. However even the most cursory look at the world of flow measurement will show that this is in fact far from the truth.

It is difficult to think of a sector of industry in which a flowmeter of one type or another does not play a part. It is therefore not surprising that the total sales of flowmeters in the world in 1988 was worth £1.1 billion. However, the value of product being measured by these meters is even more staggering. For example, in the UK alone, it was estimated that in 1988 the value of oil and gas being metered was worth £20 billion (Kinghorn 1988).

Despite the value of product being metered, the accuracy of many flowmeters is still poor in comparison to those instruments used for the measurement of other common process variables, such as temperature and pressure. For instance, many countries still specify the orifice plate as being the only flowmeter approved for the fiscal measurement of gas. Surprisingly, this device has a typical accuracy of only ±2% of full scale, so there is still room for improvement. With the value of metered products being so high, obviously any improvement in measurement accuracy can mean a large financial benefit.

It is not surprising that research is continuing to improve the understanding, and of course accuracy of flow measurement techniques. In some respects, however, the problem is becoming more rather than less difficult. Until recently, most flowmeters have been used to measure single-component materials (petrol, gas, water etc) at points of use or in production plants. However, the increasing need to use resources more efficiently and build processing plants that can be operated cost effectively, while minimizing pollution, has led to an increasing demand for multi-component flowmeters.

The development of multi-component flowmeters is both a fascinating and expanding area of measurement. Future process requirements, with their need for ever more detailed measurements, will ensure that this expansion continues.

This book is concerned with the development of a multi-component flow measurement technique, that of flow imaging, which has the potential to solve many of the above problems.

1.2 THE IMPORTANCE OF MULTI-COMPONENT FLOW MEASUREMENT

The growing demand for flowmeters capable of metering multi-component flows can be illustrated by considering two examples from the petroleum industry.

The first concerns the problem of three-component measurement. The basic features of a typical offshore production facility are shown in figure 1.1. Fluid from each well is pumped to the production platform where it undergoes preliminary processing before being shipped or piped directly to shore. The crude oil produced by an offshore reservoir usually contains both gas and water components. The flow pattern of the resulting mixture varies with the flow conditions and is generally not predictable. It is important for the operator of a production platform to know which types of fluid a well is producing. The current method of solving this problem is to separate the fluids first, and then monitor each using conventional single component flowmeters (eg orifice plate for gas, turbine meter for oil). With the space on a production platform becoming more expensive, and the development of subsea production systems increasing, the use of conventional offshore separators is becoming less desirable.

A more compact partial separation component flowmeter is shown in figure 1.2. This uses an inclined vessel to separate the liquid and gas components which can then be individually metered. The fraction of water in the oil is measured by microwave absorption. The whole system is heavy—45 tons for a 6-inch flow pipe, the auxiliary instrumentation is complex and could be difficult to maintain in hostile environments.

Therefore a particularly interesting but difficult flow measurement problem arises, namely how can the component fractions of an oil/gas/water mixture be reliably monitored without separation in a hostile environment.The qualities required for such a measurement system are that it should be:

- non-intrusive, to avoid sensor erosion and pressure drop,
- a real time measurement, to provide instantaneous feedback to the production operator,
- in-line to avoid the problem of sample representivity,
- reliable, because maintenance will be costly, and perhaps impossible in the short term,
- easily calibrated for use subsea.

A number of solutions have been proposed to this problem. One of the most promising methods currently under development uses the combination of a gamma radiation sensor and a capacitance sensor (Dykesteen and Frantzen

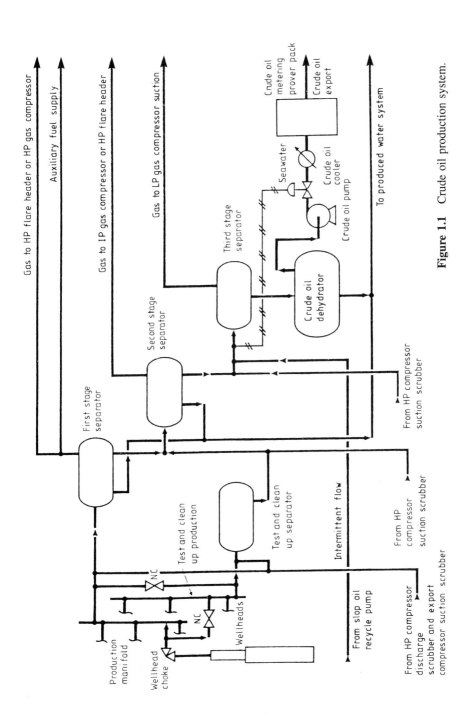

Figure 1.1   Crude oil production system.

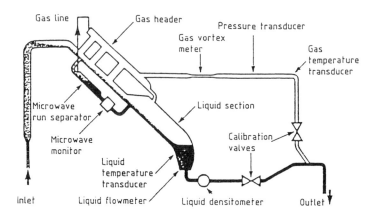

Figure 1.2   The Texaco multiphase (Volumeric flowrate) meter.

1990). The gamma radiation sensor is used to measure the mean density of the multi-component flow, while the capacitance sensor is used to determine the mean permittivity of the flow. These two independent measurements are then used to determine the component ratio of the flowing oil/water/gas mixture. Results so far show that this system is capable of determining the composition of a well mixed flow with an accuracy of better than ±3% of full scale. While there are still many problems to solve, such as how to deal with changes in flow regime, ten years ago such a flow measurement system would have been inconceivable.

A second illustration of the need for multi-component flow meters can be found by considering a later stage of the oil production process. Even after separation, the water content of a ship's crude oil cargo may lie in the range 0–5%. This may not seem a great deal, but an error of only 0.5% in measuring the water content of a 250 000 ton oil cargo would result in a discrepancy of approximately $100 000 (1990 prices). An accurate two-component measurement is therefore required; however most currently used methods have limitations.

The usual method of measuring water content today is by use of an isokinetic sampling system. This basically consists of a probe inserted into the pipeline through which samples of the flow are drawn for off-line laboratory analysis. This technique presents a number of problems, the most serious being the ability of the system to take representative samples and the time required to obtain the measurement.

Recently a number of organizations have investigated on-line water content measurement using capacitance sensors (Gimson 1989). The measurement principle has attracted a great deal of interest because it has a fast dynamic response and offers the possibility of constructing a non-intrusive sensor which can be used on-line.

The capacitance based water content meters produced so far are an improvement on isokinetic samplers in that they produce instantaneous measurements. However, their calibration is dependent on flow regime, temperature and whether the water or oil is the dispersed component in the flow.

The above examples show how rapid advances in process production techniques have encouraged the development of specialized multi-component flowmeters. It should be borne in mind however that commercial multi-component flowmeters are still rare and that the measurement techniques discussed in section 1.5 still dominate.

1.3   TERMINOLOGY

In this book we consider multi-component flows to be those where solid, liquid or gas components exist in a separated form. In many publications the term 'multi-phase flow' is used instead of 'multi-component flow', although multi-phase flows should strictly only be those composed of separate phases such as a liquid and gas. An oil–water flow is therefore by definition a two-component flow while a gas–oil flow may be classified either as a two-phase or multi-component flow. A distinct feature of multiphase flows is that they include interfaces between two materials (gas/liquid etc) which change in time and space.

Although this book concentrates on the problem of two-component flow measurement, that is liquid–liquid, gas–liquid, liquid–solid and gas–solid flows, many of the principles and techniques described are applicable to flows with more than two components.

1.4   TWO-COMPONENT FLOW MEASUREMENT: THE BASIC PROBLEM

Two-component flow measurement is not an easy matter, the first problem being to decide what parameter actually needs to be measured. When used by itself, the term 'flow measurement' is rather vague and can often mean different things to different people. For instance, one user may be interested in measuring a mixture's volumetric flow rate, while in another application the mass flow rate will be required. Since both of these may be called 'flow measurements', care should be taken when specifying the actual measurement required.

A pseudo-homogeneous fluid is probably the easiest two-component mixture to measure. In this particular case one component is finely and evenly dispersed in the other, and so the mixture can be treated as a single fluid that obeys all the usual equations of single-component flow.

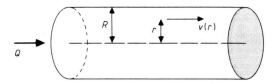

Figure 1.3   Measurement of pseudo-homogeneous flow.

Referring to the pseudo-homogeneous flow shown in figure 1.3.   The volumetric flow rate Q of the mixture can be calculated using:

$$Q = \int_0^R 2\pi r v(r) \, \delta r \qquad (1.1)$$

where $v(r)$ is the velocity of the fluid at radius r. This assumes that both components are travelling at the same velocity, and that the velocity profile is the same across all radii.

Although a velocity profile exists across the cross-sectional area of the pipe, it is usually not practical to determine Q by measuring v across the whole pipe cross-section. For a single-component flow, this problem is overcome by assuming a velocity profile across the pipe, and then using the cross-sectional average of this profile $\bar{v}$, to calculate the volumetric flow rate. A similar approach can be used in the case of a pseudo-homogeneous flow, although the characterization of a two-component flow as either laminar or turbulent in order to determine the appropriate flow velocity profile required is more difficult than for single-component flows.

Equation (1.1) can therefore be written as:

$$Q = \pi R^2 \bar{v} = A\bar{v}. \qquad (1.2)$$

The mass flow rate M of the mixture can be calculated from

$$M = \rho_m A\bar{v} \qquad (1.3)$$

where ρ_m is the density of the mixture. Equation (1.3) assumes that the two components are so well mixed that any portion of the flow will have the same density as any other portion. This mixture density will lie somewhere between the densities of the individual components (ρ_1 and ρ_2) and depend on the component volume ratio α of the mixture. Now assuming that both components are a fluid, or if particulates that the effect of particle packing is not significant: $\rho_m = \rho_1(1 - \alpha) + \rho_2\alpha$.

The usefulness of equation (1.3) will depend on whether or not the flow can be classified as pseudo-homogeneous. The homogeneity of the flowing mixture will depend on the degree of turbulence present, the density of each component and the component ratio of the mixture. The less homogeneous the mixture the greater will be the measurement error if equation (1.3) is used to calculate mass flow rate.

Unfortunately in many cases the flow components are neither well mixed nor moving at the same velocity. For example, in liquid–gas flows, groups of bubbles can be followed by a large slug of gas (see chapter 2). Since they experience different frictional, surface tension and gravitational forces, bubbles and gas slugs will move at different velocities to the main liquid phase, and also to each other. In such circumstances, the use of mixture density ρ_m and mixture velocity $\bar{v}$, as in equations (1.2) and (1.3) becomes less valid.

To add to these problems, in many industrial flow measurement applications, measurements of the individual component mass flow rates are required rather than just the total mixture flow rate.

The mass flow rate M_1 of component 1, and the mass flow rate M_2 of component 2 of the two-component mixture can be calculated from:

$$M_1 = \rho_1 A \bar{v}_1 \alpha \tag{1.4}$$

and

$$M_2 = \rho_2 A \bar{v}_2 (1 - \alpha) \tag{1.5}$$

where α is the fraction of the pipe cross-section occupied by component 1, and ρ_1, ρ_2 are the individual component densities.

Although equations (1.4) and (1.5) have an advantage over (1.3) in that they may be used for flows where the components are not well mixed, three independent measurements are now required to determine M_1 and M_2 (it is reasonable to assume that A, ρ_1 and ρ_2 are known). There are still situations however where even equations (1.4) and (1.5) will be unsuitable.

For example, consider a horizontal flow of a liquid–solid mixture, where M_1 is the solid and M_2 is the liquid. It is not uncommon, with this type of two-component flow, for some of the solids to settle on the bottom of the pipe in a static or sliding bed. In these circumstances, the measured solids velocity will be that of the solids suspended in the liquid phase, while the measured void fraction α will include the static solids bed. Hence the solid mass flow obtained from equation (1.4) will be erroneously high.

Despite the difficulties outlined above, if care is taken over the choice of application, two-component flows can often be successfully metered using a combination of measurement techniques.

1.5 CONVENTIONAL METHODS OF TWO-COMPONENT FLOW MEASUREMENT

Both the volumetric flow rate and mass flow rate of a two-component mixture can be determined in two main ways—either from a direct measurement or an inferential measurement.

Taking mass flow rate as an example: a direct mass flowmeter has a sensing element that reacts directly to the mass flow of the mixture through the instrument. An inferential mass flowmeter, however, measures both the instantaneous velocity and concentration of each component. These measurands and the component densities are then used to calculate the individual component and the total mixture mass flow rates (it is assumed that the density of each component is known or can be easily determined).

As was said earlier, two-component flows may be classified as gas–liquid, gas–solid, liquid–liquid and liquid–solid. Each presents its own measurement problems, and it would be impractical to discuss in detail all possible metering methods for these flow types in an introductory text such as this. The methods most commonly used for inferential and direct measurements are summarized in table 1.1, which shows whether the technique is specifically sensitive to the dispersed component, the carrier or to both. Brief details of techniques suitable for measuring the velocity of the carrier and dispersed component of a mixture are given in section 1.6. In addition, methods for component concentration measurement are described (section 1.7). Finally, the Coriolis flowmeter, currently the most popular instrument for direct mass flow measurement of two-component flows is discussed in section 1.8.

For further details of measurement techniques specifically intended for liquid–gas flows, see the reference book by Hetsroni (1982). Gas–solid and liquid–

Table 1.1 Two-phase flow measurement techniques (D dispersed; C carrier; B both)

Carrier phase	liquid	liquid	liquid	gas	gas
Dispersed phase	liquid	gas	solid	liquid	solid
Velocity measurement technique					
Laser Doppler	D	D	D	D	D
Microwave Doppler	D	D	D	D	D
Ultrasound Doppler	D	D	D	D	D
Injected tracer	C	B	B	B	B
Nuclear magnetic resonance	B	C	C	D	
Pulsed neutron activation	B	C	C	D	C
Cross-correlation	D	D	D	D	D
Concentration measurement techniques					
Radioactive attenuation	D	D	D	D	D
Capacitance	D	D	D	D	D
Ultrasound absorption	D	D	D	D	D
Optical absorption	D	D	D	D	D
Mass flow measurement					
Coriolis	B	B	B		

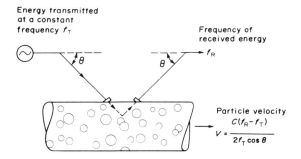

Figure 1.4 Doppler velocity meter.

liquid flows are covered in publications by Woodhead *et al* (1990) and Miller (1983), while for liquid–solid (or slurry) flows, the booklet by Baker and Hemp (1981) can be recommended.

1.6   COMPONENT VELOCITY MEASUREMENT

In some cases it may be possible to use conventional single-component flowmeters such as orifice plates, venturi meters and turbine meters to measure component velocity. The principles of these instruments are described in many standard flow measurement books e.g. Miller (1983). However, in general these instruments are only usable with two-component flows when the dispersed component is a small quantity of gas or liquid, or a very small quantity of nonabrasive and nonblocking solid. Even in these circumstances it is often difficult to predict the accuracy of the measurement with any certainty (Pursley and Paton 1985).

Velocity measurement techniques more appropriate for two-component flows fall into two main groups: Doppler and interference fringe methods and tracer methods.

1.6.1   Doppler and interference fringe methods

The Doppler shift principle can be used as the basis of a dispersed component velocity meter (figure 1.4). Energy is transmitted from a source at a constant frequency f_T and directed into the flow being monitored. Some of this energy will be reflected by discontinuities presented by the dispersed component in the flow and be received at a frequency f_R. The difference in frequency between the transmitted and received signals is related to the velocity v of the conveyed component thus:

$$f_R - f_T = \frac{2vf_T \cos\theta}{c} \tag{1.6}$$

where c is the velocity of the transmitted energy and θ is the angle made by the transmitted energy beam to the flow. As this equation shows, there is a linear relationship between the dispersed component velocity and the shift in frequency of the received signal. In addition, since θ, f_T and c are known, a flowmeter based around this principle should not require on-line calibration.

Doppler flowmeters for use with two-component mixtures can be constructed using lasers, microwaves or ultrasound as the energy source.

Laser methods. The laser velocity meter may be used in a number of different configurations (Bopp *et al* 1990), but the two most commonly used are the differential method and the reference beam method. In the differential method, figure 1.5, a laser beam is split into two parallel rays which are transmitted through a lens which causes them to cross at a point in the flow where the velocity is to be measured. The resulting intersection volume of these two rays consists of a series of interference fringes. Particles passing through these fringes will therefore cross zones of alternately high and low intensity illumination. Light scattered by these particles is detected using a photomultiplier which is focused onto the intersection volume. The amplitude modulation of the light changes frequency with particle velocity. The frequency of the detected signal is related to the velocity of the dispersed component. A limitation of the differential method is that it is not suitable for use with flows of large particles, since the signal-to-noise ratio of the detected signal is dependent on the ratio of dispersed component size to the width of fringes in the measuring volume.

For dispersed components of diameter larger than the beam, the reference beam method of operation is preferred. A single beam is directed at the point in the flow which is to be monitored. Light scattered from particles passing through this measuring point is mixed with a reference beam and the shift in frequency between the transmitted and the reflected light is detected by a photomultiplier.

Irrespective of whether the differential method or reference beam method is used, a laser can be used for point velocity measurements in the range 0.1 mm s^{-1} to 100 m s^{-1} with an accuracy claimed to be better than approximately 0.5% of measured value, and no calibration is required. Although commercially available, and an excellent research instrument, laser velocity meters are currently too expensive and fragile for routine velocity measurements in industrial plants.

Microwave methods. Several configurations of microwave Doppler velocimeter are possible, but the two most commonly used are the bistatic method and the monostatic method. The basic features of the bistatic configuration are shown in figure 1.6. A microwave signal is directed into the flow from a horn antenna through a dielectric window. The transmitted signal is scattered by discontinuities in the flow and the reflected signal is detected by a separate receiver and mixed with the transmitted signal to obtain the Doppler frequency

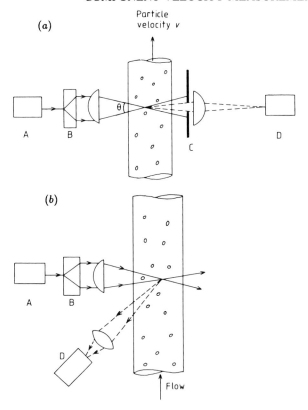

Figure 1.5   (*a*) The differential laser Doppler velocimeter in the forward scatter mode. A, laser source of the wavelength λ; B, beam splitter; C, aperture; D, photomultiplier receiving light, the intensity of which is modulated at a frequency f. (*b*) The differential laser Doppler velocimeter in the backscatter mode.

shift. At a typical X-band transmission frequency of 10.69 GHz this frequency shift is approximately 706 Hz for a 10 m s^{-1} change in velocity. The measuring volume of this configuration is defined by the overlapping beams of the transmitting and receiving antenna.

The monostatic configuration of the microwave Doppler flowmeter uses a transceiver rather than a separate transmitter and receiver. In order that a single horn can be used, isolation must be provided between transmitted and received signals. This can be achieved using a ferrite circulator. Although the monostatic configuration has a larger measuring volume than the bistatic configuration, it does have the advantage of needing only one antenna, and it is therefore easier to instal and operate.

Whichever of these two configurations is used, the detected signal will consist of not just one frequency but will be composed of a number of different frequencies. This spread of frequencies is the result of the large measuring

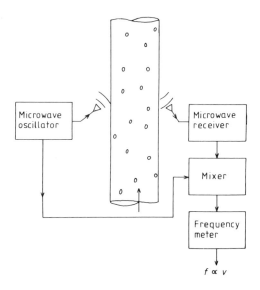

Figure 1.6   Microwave Doppler velocity meter: bistatic configuration.

volume used, which enables a large number of conveyed discontinuities (many of which will be travelling at different velocities) to reflect signals back to the receiver at the same time. Although the average frequency of the signal will bear some relationship to the average bulk velocity of the conveyed component, the possibility arises of preferential or spurious reflections leading to a bias on the reading which is not readily assessable. This is a recognized problem with many Doppler radar systems (Stuchley *et al* 1977).

Although the spatial resolution of the microwave Doppler meter is poor, it can be used with non-transparent liquids (a situation in which the more accurate and expensive laser-based systems fail). Since, in principle, the electronic components of the instrument can be mounted remotely from the conveying pipe, it may be used for the routine measurement of average conveyed component velocity in hostile and inaccessible situations. Despite these advantages, commercial microwave Doppler flowmeters are uncommon. However, with the continued development of semiconductor microwave devices this situation may well change in the future.

Ultrasonic methods. Several ultrasonic Doppler flowmeters are commercially available for use with liquid carrier systems. As with the microwave version, many configurations of ultrasonic Doppler meter are possible (Lynnworth 1989). Usually, a fixed-frequency ultrasonic signal is directed into the flow from a transmitter (normally a piezoelectric transducer) and reflected back to a receiver by discontinuities in the flow. The received signal is mixed with the transmitted carrier to produce a signal which includes the carrier modulated by the Doppler frequency. A bandpass filter can be used to remove unwanted noise and a

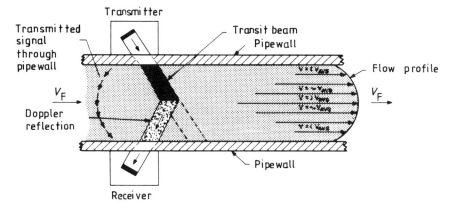

Figure 1.7 Principle of the Doppler-effect ultrasonic flowmeter with separated opposite-side dual transducers.

demodulator used to extract the Doppler frequency (figure 1.7).

For liquids, the transmitted carrier frequency may lie in the range from 500 kHz up to a few MHz. At 500 kHz the dispersed component must have a minimum diameter of approximately 50 μm in order to reflect ultrasound back to the receiver. Increasing the operating frequency will allow the detection of smaller scattering centres, but at the cost of reducing the penetration of the transmitted signal. As a result, the Doppler frequency will be strongly related to the velocity of discontinuities flowing near the pipe wall.

As with microwave versions, the ultrasonic Doppler meter is sensitive to changes in flow profile and the spatial distribution of discontinuities in the flow. Since these are difficult to predict, measurement errors due to these variations can be most easily reduced by calibrating the meter on-line.

The velocity of sound in a fluid is related to both the temperature and density of that fluid, and changes in temperature or density of the conveyed carrier will result in a shift in the calibration of an ultrasonic Doppler meter. This problem can be minimized by measuring the temperature of the fluid being metered, and then adjusting 'c' in equation (1.6) as the fluid temperature changes. Automatic temperature compensation schemes may be easily implemented using techniques such as digital look-up tables, or by solving an appropriate polynomial equation.

Ultrasonic Doppler meters may be 'clamp-on', but changes in calibration will occur unless the angles at which the ultrasound beams enter and leave the flow are kept constant. As a result, meters which are permanently bonded to a pipe will have a better repeatability.

The accuracy of most commercial ultrasonic Doppler meters is poor (typically 5–10% of full scale), although this can be improved by calibrating the meter on-line. If care is taken with installation, a repeatability of approximately 1% of full scale can be achieved on pipe diameters of less than 5 cm.

Attempts to extend this technique to gas carrier systems have met with only limited success. The main difficulty encountered is that of obtaining an efficient

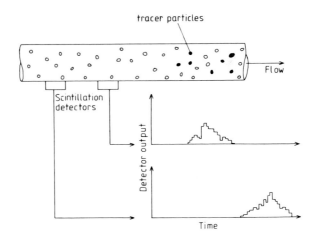

Figure 1.8 Velocity measurement using radioactive tracer particles.

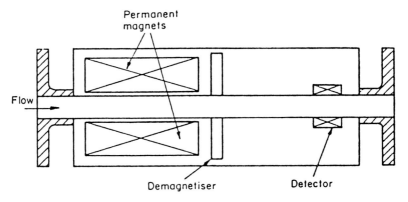

Figure 1.9 Simplified schematic arrangement of the nuclear-magnetic-resonance flowmeter (From Hayward A T J 1979 *Flowmeters* (Macmillan Press)).

energy coupling into the gas flow from the ultrasonic transmitter. In order that this may be achieved, source frequencies in the region of 40 kHz should be used. However, beam angles are wide in these circumstances, leading to problems with the definition of the measuring volume.

1.6.2 Tracer methods

The tracer method of velocity measurement is fundamental, and two variations of this technique can be used to measure component velocity. Either recognizable markers can be injected into the flowing stream and their progress timed between two points, or signals due to natural disturbances can be detected at two points in the flow stream and the time difference determined by cross-correlation techniques.

Intrusive injection. For injected tracer methods to be suitable for velocity measurement in pipelines, it is desirable that two basic conditions should be met by the tracer used. First, the passing of the tracer should be distinguishable from that of the rest of the conveyed phase by a sensor placed on the outside of the pipe. Second, the tracer and component being measured should have the same density and size distribution, otherwise the two will travel at different velocities.

One way of meeting these requirements is by using radioactive particles as a tracer. If a pulse of these radioactive particles is injected into the flowing stream, the passing of the resultant cloud of radiation can be detected at two points downstream by scintillation detectors, and the velocity of the flow thus determined. The first detector should be placed a sufficient distance downstream of the injection point to enable the tracer particles to accelerate to the main stream velocity and also to allow the tracer particles to be mixed across the full diameter of pipe. Unfortunately, this acceleration distance also allows the tracer particles to become dispersed in the direction of travel. The result of this dispersion is that each scintillation detector outputs a burst of pulses rather than a sharply defined pulse as the tracer cloud passes by it (figure 1.8). Measurements between the 'centre of gravity' of each pulse are found in most cases to give the most repeatable results.

Although injected tracer methods can be very successful, they do require the service of a team of specialists who are well versed in their use and in the necessary safety aspects. This makes them expensive and tends to limit their application to in-depth research investigations on process plants.

Potential users of radioactive tracer methods are strongly advised to contact their local radiological applications and safety organization (e.g., UKAEA Harwell in the UK, Argonne National Laboratory in the USA, CEA in France), who should be able to advise on the correct and safe use of the method.

Non-intrusive injection. The nuclear magnetic resonance The nuclear magnetic resonance (NMR) (figure 1.9) flowmeter is basically a tracer method of velocity measurement. The spin-state magnetic resonance is created by a combination of radio frequency and permanent magnet fields. The RF field is time-coded, so that the delay time at a suitable spin state receiver in a downstream position can be measured. This delay time is inversely proportional to the flow velocity (Genthe 1974).

NMR flowmeters possess many ideal qualities. They have been successfully used in multi-component applications such as the measurement of a coal–oil slurry fuelling a blast furnace and for single-component fluids such as fuel oils and salt water.

Even though NMR meters are now commercially available they are expensive. A great deal more will need to be known about their performance before they become accepted by industrial users.

The pulsed neutron activation technique. Like the NMR flowmeter, the

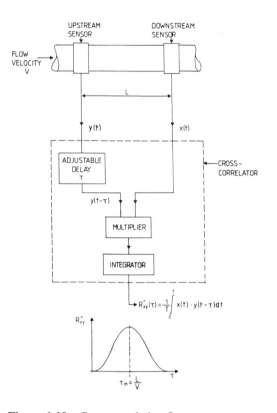

Figure 1.10   Cross-correlation flow measurement.

pulsed neutron activation (PNA) technique is also a tracer method of velocity measurement. A high-energy source placed on the outside of the pipe directs neutrons into the flow. This activates oxygen atoms in the fluid molecules to produce the radioisotope N_{16} which has a half-life of 7.12 s. The pulses of radiation are detected at two points downstream using scintillation detectors and the velocity of the flow can therefore be determined. Each scintillation detector outputs a burst of pulses rather than a sharply defined pulse as the 'radioactive cloud' passes. As with intrusive injection methods, the interpretation of the results is difficult and similar analysis techniques are required.

PNA measurement systems are not yet readily available commercially and, as with all radiation methods, stringent safety precautions must be taken. However, recent work has shown that by measuring the radiation count rate downstream of the activation pulse as a function of time, the PNA technique can be used to measure the mass flow rate as well as the velocity of multi-component flows. It is probable that this method of measurement may be increasingly used in the future.

The cross-correlation method. The principle of the cross-correlation method of flow measurement is shown in figure 1.10. Two sensors are used to monitor the flow, one being positioned downstream of the other. These sensors detect variations in some naturally occurring property of the fluid, such as density or temperature. The output signal of each sensor is therefore modulated by spatial and temporal variations in the detected property of the passing fluid, in an apparently random manner. However, assuming that the pattern of the conveyed discontinuity remains unaltered as it travels between the sensors, then the output signal generated by the downstream sensor will be a time-delayed replica of the upstream sensor's output. In most practical cases the discontinuity pattern will not remain frozen, but a recognizable part will be conveyed between the two sensors.

The time delay between the output signals of the two sensors can be found by computing the cross-correlation function of their time records $x(t)$ and $y(t)$ over a measurement period T. This cross-correlation function is given by:

$$\hat{R}_{x,y}(\tau) = \frac{1}{T} \int_0^T x(t)y(t - \tau)\, dt \qquad (1.7)$$

where $\hat{R}_{xy}(\tau)$ is the value of the cross-correlation function when the upstream signal $y(t)$ has been delayed by a time τ.

The transit time of the flow between the two sensors is found by observing the time-lag τ_m at which the cross-correlation function is a maximum. Since the distance between the sensors is known then, assuming that the detected tracer is moving at the same velocity as the fluid, the velocity of the flow can be found from:

$$v = \frac{L}{\tau_m} \qquad (1.8)$$

where L is the sensor spacing.

The volumetric flow rate Q then can be calculated from:

$$Q = kvA \qquad (1.9)$$

where A is the cross-sectional area of the pipe and k is a meter factor to allow for the non-uniform velocity profile.

Cross-correlation flowmeters are ideally suited to the velocity measurement of multi-component flows, and also to the problem of obtaining reliable measurements in inaccessible locations. As an example of the latter, the measurement of coolant flow in nuclear reactor heat exchanger tubes was one of the first applications of the cross-correlation technique to flowmetering.

This technique may be implemented using a wide variety of sensors (some very simple indeed), such as conductivity, capacitance, ultrasonic and temperature (Beck and Plaskowski 1987). The only requirement is that the sensors are able to detect some random fluctuation in a physical property of the flow, and transform this into a corresponding electrical signal which can be cross-correlated to determine the flow velocity.

Since transit time alone is measured, the sensors may be simple and sensor calibration is generally not required (only one initial calibration is needed to determine the meter factor, k, in equation (1.9)). The cross-correlator for measuring the transit time of the disturbances is now a relatively inexpensive device because of the low cost of digital electronic processing.

Although still commercially in its infancy, the future for the cross-correlation flowmeter looks promising. It is now no longer just a research technique being used in one-off applications, but now, with the introduction of commercial cross-correlation flowmeters, is finding increasing use in the industrial sector.

1.7   COMPONENT CONCENTRATION MEASUREMENT

The two most popularly used methods for the measurement of component concentration in two-component flows are based on radiation and capacitance techniques. These methods will be outlined and a few alternatives mentioned.

1.7.1   Radioactive attenuation methods

Measurement methods based on the attenuation of radiation have been developed extensively during the last 15 years. With all these methods a source of radiation (commonly γ-rays and x-rays) is placed on one side of the pipe and the attenuation of the beam of radiation after it had passed through the flow is measured.

This attenuation, I/I_0, is given by the equation:

$$I = I_0 \exp(-\mu D) \qquad (1.10)$$

where I_0 is the intensity of incident radiation, I is the intensity of transmitted radiation, μ is the linear absorption coefficient and D is the path length across the pipe.

In a two-component mixture of, say, gas and liquid this expression becomes:

$$I = I_0 \exp[-(\alpha \mu_G D + 1 - \alpha \mu_L D)] \qquad (1.11)$$

where μ_G is the linear absorption coefficient of gas, μ_L the linear absorption coefficient of liquid and α the volumetric fraction of gas. The expression can be used to determine α if μ_G, μ_L and D are known.

If these parameters are not known it is possible to use the instrument by effectively calibrating on line thus:

$$\alpha = \frac{\ln I - \ln I_L}{\ln I_G - \ln I_L} \qquad (1.12)$$

where I_L is the received radiation with the pipe full of liquid and I_G the received radiation with the pipe full of gas.

Although in concept radiation methods of component concentration measurement are simple and elegant, in practice a number of difficulties have to be overcome.

First, because of the statistical nature of the measurement, there is a compromise between measurement time and accuracy. The greater the accuracy required, the longer will be the measurement period (Lassahn *et al* 1979). Of course, stronger sources of radiation will reduce the required measurement period, but in these circumstances far greater safety precautions have to be taken.

The second problem with single beam radiation methods is that they are flow-regime dependent. The line average of component concentration calculated from equation (1.11) will only be representative of the complete flow cross section if the components are homogeneously mixed.

Single beam systems have a typical accuracy of ±5–10% of full scale, but this depends on the accuracy of calibration, the variation of flow regime and the signal-to-noise ratio. In certain cases the size and weight of these systems can be a disadvantage, for instance a void fraction meter for a 100 mm diameter pipe weighs about 60 kg.

In situations where the flow regime is unknown or unstable, a single source producing several collimated beams may be used, each beam requiring a separate detector. These multibeam densitometers are complex and expensive and, as yet, very few are commercially available. However, component concentration can be measured with an accuracy of a few percent using these devices and so they are being increasingly used in multi-component mixture research applications.

1.7.2 Capacitance methods

Capacitance methods of component concentration measurement are based on the principle that the electrical capacitance between two electrodes, through which a multi-component mixture is flowing, is dependent on the component concentrations. This method of concentration measurement has attracted a great deal of interest, because, like radiation methods, the principle behind the measurement is simple. Moreover, it has advantages over radiation methods in that an almost instantaneous dynamic response can be achieved and there are no radiation hazards.

Measurements of change in static capacitance can be successfully used to measure component concentration when the concentration of the discontinuous component is high. This method is also useful for measuring low concentrations of water because of the high relative permittivity. However, design of the reliable electronic circuitry to measure the capacitance of the sensor is usually not easy, since in most circumstances the capacitance changes to be measured are small (Huang 1986).

At lower concentrations a capacitance 'flow noise' technique may be used. The flow noise is caused by the variations about the mean particle density caused by the essential turbulence in multi-component flow. The rms value of this flow noise is then functionally related to component concentration. This type of meter does require specific calibration for the particular type of discontinuous component but it has the inherent advantage of being completely free of zero drift (Green 1981, Green and Taylor 1986).

The main disadvantage of both these simple capacitance methods of component concentration measurement is that their calibration is dependent on flow regime. Two recent developments have used very different approaches in an attempt to produce an instrument which would be independent of flow regime. Both of these are now commercially available.

The 'rotating' field sensor consists of three pairs of electrodes spaced around the outside of the pipe in order to produce a spatial averaging. As shown in figure 1.11, each of these electrode pairs is excited by a 5 kHz wave such that each pair is 120° out of phase with the others. This type of sensor has been used successfully to meter traditionally difficult two-component flows such as pneumatically conveyed coal particles (Mathur and Klinzing 1984).

An alternative to the rotating field sensor uses a helical construction to provide spatial averaging. The helical sensor, although of complex mechanical construction, has a reasonably linear calibration which is flow-regime independent. Commercial helical sensors have been developed for the on-line measurement of water in crude oil, and a typical accuracy of ±1% of full scale over a water fraction range of 0–80% is claimed (Hammer *et al* 1989).

1.7.3 Alternative methods of concentration measurement

New methods for component concentration measurement of multi-component flows continue to be invented. Many of these methods are initially designed to overcome a measurement problem in a particular research project, but usually are never subsequently developed commercially. A few alternatives to radiation and capacitance techniques which may be worth considering in some applications will now be described.

The absorption of ultrasound can be used to measure comparatively high concentrations of the second component in a liquid carrier. However, the uncertainties and variations in the complex propagation of ultrasound make a direct absorption technique unsuitable for low conveyed component concentrations, say below 5%. At lower concentrations the modulation of an ultrasound beam by particle scattering and absorption can be used to indicate solids concentration. This principle has been used to measure the concentration of solids in sewage plant liquors (Balachandran and Beck 1980).

In general, ultrasonic techniques in gases suffer from the problems of relatively poor transmission. However, some success has been obtained in measuring the concentration of solids in gas using an absorption technique.

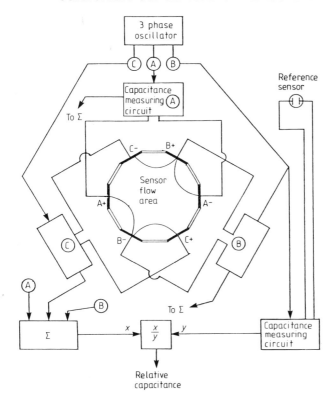

Figure 1.11   The Auburn 'rotating field' capacitance sensor.

Optical techniques are accurate only at comparatively low concentrations of the discontinuous component because any optical effects become grossly nonlinear at high concentrations due to light obscuration by overlapping particles.   Optical absorption techniques have been successfully used for measuring the solids concentration in effluent stack gases, and ingenious systems using laminar flow windows have been used to reduce problems due to solids deposition on transparent sections. Optical absorption systems inevitably suffer from gross errors if the particle size deviates from the calibration value. Instruments relying on multiple-angle scattered light have been devised which use a microcomputer to calculate the particle size distribution and concentration, but these are rather complex for process pipeline use (Harris *et al* 1976).

Optical absorption and scattering instruments are sometimes used for multi-component liquid systems, but care must be taken to keep the optical surfaces clean. An ingenious instrument has been developed for measuring the oil content of an ocean-going tanker flushing water (Pitt *et al* 1985).   This uses fibre-optic light guides to transmit and receive the measuring light beam, so that all electrical components can be remote from the potentially flammable atmosphere near the sampling cell, and incorporates a water jet washer to periodically clean the optical surfaces.

Component concentration in liquids may be determined from direct measurement of density. Techniques for density measurement are described in most standard textbooks on process instrumentation, see for example Considine (1985). In general density measurement will only be satisfactory for determining component concentration if two conditions are met: (a) the density of each component is constant, and (b) the component density ratio is high.

Among the available density meters, the gravimetric types are suitable for multi-component flows. The use of vibrational-type density meters is often inadvisable because the non-visco-elastic behaviour of multi-component mixtures can cause considerable errors.

1.8   DIRECT MASS FLOW MEASUREMENT

There are a number of methods for the direct mass flow measurement of fluids, but the instrument with the fastest growing reputation is without doubt the Coriolis flowmeter. A paper by Medlock and Furness (1990) gives a good review of all methods of mass flow measurement. The following sections will only concentrate on the Coriolis flowmeter.

1.8.1   The basic principle of Coriolis mass flow measurement

The mass flow rate of a fluid (or in some circumstances a mixture) can be measured using the Coriolis principle. Referring to figure 1.12, here the flow is moving at a velocity v, while the tube is rotating with angular velocity ω about a fixed point P. Consider the zone of fluid denoted δx in figure 1.12. This zone will experience an acceleration 'a' in the direction perpendicular to the wall of the tube and in the plane of rotation. The acceleration arises from two causes. Firstly, as the zone moves away from point P at constant fluid velocity v its tangential speed will increase in proportion to the distance x from point P. Secondly, as the tube rotates the direction of fluid matter changes, giving an additional tangential acceleration. The force required to cause this acceleration 'a', must result in an equal and opposite force δF on the wall of the tube. This force δF is a Coriolis force and is given by:

$$\delta F = 2\omega v\rho A\delta x \tag{1.13}$$

where ρ is the fluid density and A the cross-sectional area of the tube.

The mass flow rate of the fluid is given by:

$$M = v\rho A \tag{1.14}$$

and substituting in equation (1.13) gives:

$$\delta F = 2\omega M \, \mathrm{d}x. \tag{1.15}$$

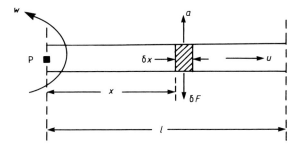

Figure 1.12   Coriolis force δF experienced by shaded zone in fluid moving at velocity u.

The Coriolis force F on the full length of the tube in figure 1.12 is given by:

$$F = \int_{x=0}^{l} 2\omega M \, \mathrm{d}x = 2\omega M l. \tag{1.16}$$

Since ω and l are known, then from equation (1.16), the mass flow rate of the fluid M is proportional to the Coriolis force F. This is the simple principle on which all Coriolis mass flowmeters are based.

1.8.2   Practical considerations

The rotating tube arrangement shown in figure 1.12 is not used in practice because it would require complicated and potentially unreliable mechanical seals. Therefore, the commercial instruments available (Medlock and Furness 1990) operate by vibrating the measurement tube instead of rotating it.

The Coriolis force equations above apply equally well to vibrational and rotational movements. In the case of vibrational systems the Coriolis force results in flexure of the tube. Since these flexures are small, reliable and accurate methods of measuring small displacements have been developed.

The first commercial Coriolis flowmeter used a U-shaped tube, but now many configurations are available from over a dozen manufacturers, including dual loops and a straight through design (Cascetta *et al* 1989). Each design has its own merits, with factors such as accuracy, repeatability, linearity and pressure drop, varying from design to design.

Whichever design is used, the Coriolis flowmeter has an accuracy of $\pm 1\%$ of reading (or better). The measurement range varies with manufacturer, 10:1 being typical. Similarly, pipe sizes vary, and although a $6''$ flowmeter is available from one manufacturer it is heavy and cumbersome to instal.

1.8.3   Performance with two-component flows

While the Coriolis flowmeter is used to measure total mass flow rate, it cannot in itself measure the mass flow rate of individual components. If, however, the two-component flow being metered is pseudo-homogeneous, a measurement accuracy approaching that for single-component flows may be possible.

For example, reliable measurements of liquid/gas flows are possible if the gas component is well distributed and less than 10% of the flow. However, if the gas component is so large that slugs are formed in the flow then measurement problems will occur. This is because the Coriolis flow meter depends on the material inside the metering tube following the motion of the metering tube precisely. This must occur with homogeneous fluids but not necessarily with non-homogeneous mixtures. This phenomenon is likely to reduce the feasibility of using Coriolis flowmeters on two-component flows where the phases are well separated.

With liquid/solid flows (slurries), the user has to compromise between avoiding particle dropout and avoiding excessive fluid velocities which would result in accelerated wear of the flow tube.

By comparison with most other commercial flowmeters, the Coriolis flowmeter is still very new. Not surprisingly therefore, much has still to be learnt - about the performance of this type of instrument, particularly when used with multi-component flows (Furness 1990). However, the Coriolis flowmeter has already gained a good reputation for being able to accurately meter traditionally difficult products such as ice cream and syrup. It has a growing reputation for use with two-component flows, and this looks likely to continue.

1.9   THE FLOW IMAGING METHOD OF TWO-COMPONENT FLOW MEASUREMENT

If they are applied with care, the techniques described in the previous sections may be used to obtain adequate measurements in many two-component flow applications. However, if the components in the flow are moving at different velocities, and the flow regime is non-homogenous and unstable, then the reliability of the measurement will often be uncertain. As a result, measurement errors will usually be variable and large.

In these circumstances if accurate measurements of volumetric or mass flow rate are required, then both the density and velocity distribution across the cross-section of the pipe must be known. The most practical way of obtaining this information is by using the flow imaging (or tomography) technique. An outline diagram of an electrical flow tomography system is shown in figure 1.13, a description of the subsystems in the figure is given in section 1.10.

Most people today associate tomography with the complex systems used to obtain images of internal parts of the human body. Tomography systems, however, have applications in other areas of science and technology, one of these being multi-component flow measurement.

In medical imaging the sensing system must be moved axially along the body to obtain three-dimensional images. However, in flow imaging the flow field

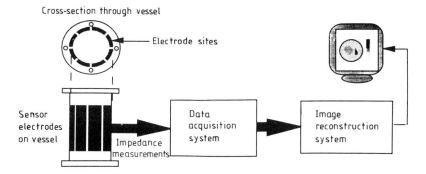

Figure 1.13   Schematic block diagram of a flow imaging system.

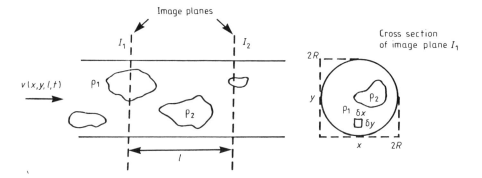

Figure 1.14   Flow imaging for measurement of non-homogeneous flow.

moves along the pipe so that a single image plane is sufficient to characterise the flow. In addition, the axial velocity of each of the components over the cross-section of the pipe must be measured, and this may be done by cross-correlation of information from two image planes spaced along the pipe axis.

Figure 1.14 shows the basic principle of the flow imaging method of two-component flow measurement. A two-component flow is moving axially between the two image planes I_1 and I_2, which are spaced sufficiently close together for there to be only a small dispersion of the flow field between the planes.

Consider a small element $\delta x \delta y$ in the cross section of the flow, located at position x, y in image plane I_1. The element may consist of either component 1 or component 2 and letting the element mass flow rate at time t be δM_1 or δM_2 (depending on whether it is component 1 or 2 we can write):

$$\delta M_1(x, y, t) = \delta x \delta y \rho_1 v(x, y, t) w_1(x, y, t) \qquad (1.17)$$

and

$$\delta M_2(x, y, t) = \delta x \delta y \rho_2 v(x, y, t) w_2(x, y, t) \qquad (1.18)$$

where v is the velocity, ρ_1 and ρ_2 are the densities of components 1 and 2 and w_1 and w_2 are binary weighting functions to denote whether the element is composed of component 1 or component 2. i.e.

$$w_1(x, y, t) = 1 \quad \text{and} \quad w_2(x, y, t) = 0 \tag{1.19}$$

where the element is composed of component 1 and

$$w_1(x, y, t) = 0 \quad \text{and} \quad w_2(x, y, t) = 1 \tag{1.20}$$

where the element is composed of component 2.

The average mass flow rates of the two components over the whole cross section of the pipe are obtained by integrating the small element flow rates in equations (1.17) and (1.18) over space and time. The result is:

$$\hat{M}_1 = \frac{1}{T} \int_{t=0}^{T} \int_{x=0}^{2R} \int_{y=0}^{2R} \rho_1 v(x, y, t) w_1(x, y, t)\, \delta x\, \delta y\, \delta t \tag{1.21}$$

and

$$\hat{M}_2 = \frac{1}{T} \int_{t=0}^{T} \int_{x=0}^{2R} \int_{y=0}^{2R} \rho_2 v(x, y, t) w_2(x, y, t)\, \delta x\, \delta y\, \delta t. \tag{1.22}$$

Equations (1.21) and (1.22) represent a complete solution to the two-component flow measurement problem. This solution requires measurement of the cross-sectional density profile $\rho(x, y, t)$ which is obtained by a tomographic imaging system, as described in the remainder of this book.

In addition, the cross-sectional velocity profile $v(x, y, t)$ is needed. If we assume that the flow cross section includes sufficiently well identified boundaries between phases, and that these boundaries move at the same speed as the nearby fluid, the velocity profile can then be obtained by cross-correlating the density information between the two image planes I_1 and I_2 in figure 1.14 to give, as described in section 1.6.2, the cross-correlation function:

$$\hat{R}_{I_1, I_2}(x, y, \tau) = \frac{1}{T} \int_0^T \rho_{I_2}(x, y, t) \rho_{I_1}(x, y, (t - \tau))\, \delta t. \tag{1.23}$$

From equation (1.23) we compute the time delay $\tau^*(x, y, T)$ of the maximum value of the function $\hat{R}_{I_1 I_2}(x, y, \tau)$.

The velocity profile is therefore given by:

$$v(x, y, T) = \frac{L}{\tau^*(x, y, T)} \tag{1.24}$$

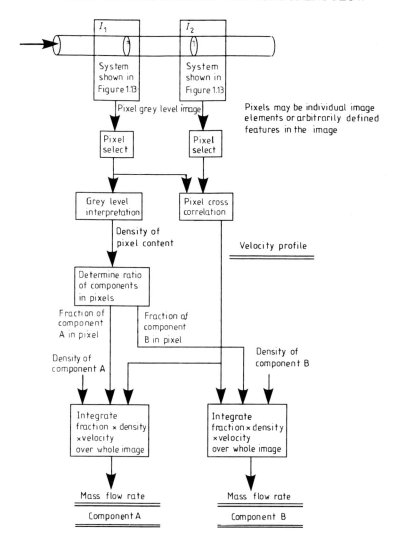

Figure 1.15   Measurement of velocity profile and component mass flow.

where L is the spacing between I_1 and I_2.

Figure 1.15 shows the system represented by equations (1.19)–(1.24) in the form of a block diagram.

Further details of the basic principles of cross-correlation velocity measurement are given in the book by Beck and Plaskowski (1987), and its application to velocity profile measurement is currently being investigated (Hayes *et al* 1992).

1.10   THE BASIC SUBSYSTEMS FOR FLOW IMAGING

Although the detailed design of a flow imaging system will depend on the process with which it is to be used and the output information required, all systems can be broken down into three basic blocks; the sensor, the sensor signal-processing and the image reconstruction and display (figure 1.13).

1.10.1   The sensor subsystem

As with all measurement systems the sensor is probably the most critical part of a flow imaging system. The manner in which the sensor interrogates the flowing mixture, and the quality of information obtained as a result of this, has a profound effect on the reliability and accuracy of the complete system.

Practical considerations place a number of constraints on the sensor. It must be compact, non-intrusive, require minimum maintenance or calibration, and, if inflammable materials are present, introduce no electrical hazards.

Sensing techniques currently under consideration for use in flow imaging systems include electrical, ultrasound, nucleonic and optical (chapter 3). Most sensors can be categorized as having either a hard or a soft sensing field. With hard field sensors, such as nucleonic and optical, the sensor field sensitivity is not influenced by the nature of the flow being imaged.

However with soft field sensors, such as capacitance and conductivity, the sensing field is altered by the phase distribution and physical properties of the mixture being imaged. Ultrasound sensors are in an intermediate category because scattering effects may interfere with the rectilinear propagation of the pressure wave.

Flow images are usually required in real-time, therefore the dynamic response of the sensors is also an important parameter. With electrical sensors this does not represent a significant problem; however, with nucleonic sensors, achieving an adequate signal-to-noise ratio requires that long integration times be used.

1.10.2   The sensor signal processing sub-system

The sensor output can be
  (i)   analogue (such as capacitance, impedance and inductance),
  (ii)   mixed analogue and digital (such as ultrasonic echo waveform and time delay),
  (iii) digital such as x-ray photons.
  This sub-system involves two major aspects.
  (i)   it converts the output of the sensor sub-system into the form required by the image reconstruction and display sub-system,
  (ii)   in many industrial situations it may need to be located at some distance from the sensor for safety or environmental protection.

1.10.3   Image reconstruction and display

The most critical part of the system software is the image reconstruction algorithm (Chapter 4). This involves the use of extended, sometimes iterative, calculations on large sets of data and inevitably is the major item of data processing cost. However, methods of image reconstruction cannot be chosen in isolation. For example the number of sensors used to image the flow should be kept to a minimum, otherwise cost and installation difficulties may escalate, and therefore the reconstruction algorithm must be capable of working with data from relatively few views of the flow. This is somewhat different from the situation in medical imaging where there may be much more freedom to move the transducer to many positions around the object to be imaged. The success of image reconstruction may depend upon adequate data pre-processing to remove the effects of discontinuities, noise, etc, in the measured data.

Another aspect of the software subsystem is that it may be used to control and periodically self-test the sensor subsystem. Simple examples would be the use of the software to switch a capacitance sensor to different measurement points around the pipe and to self-test by switching the sensor to accurately known zero and full scale reference capacitors.

1.11   THE NEED FOR A SYSTEMATIC APPROACH TO DESIGN

Flow imaging is an emergent field of technology, much aided by contemporary electronic systems and low cost data processing. The complexity of the task indicates the need for a systematic approach followed by the division of the flow imaging system into a number of subsystems. The following chapters aim to describe the scientific and engineering details of the subsystems involved and a thorough knowledge of their content should enable the reader to solve specific problems. No doubt more 'case histories' will develop in the next few years, so that in future a more 'turn-key' approach to multi-component flow measurement and flow imaging will be possible.

1.12 SUMMARY

The main points of this chapter are:
- Advances in process production techniques have led to an increasing demand for multi-component flowmeters.
- For pseudo-homogeneous flows, simple measurements of density and mean velocity can readily give the mixture mass flow rate.
- There are a wide range of instruments for velocity and concentration measurement, which, if applied with care, can be used to obtain adequate volumetric and mass flow rate measurements in many two-component applications.

- For accurate measurement of non-homogeneous flow rate, density and velocity distribution across the cross-section of the pipe must be known. The most practical way of obtaining this information is by using the flow imaging technique.
- Flow imaging systems are complex and although they can be broken down into smaller subsystems, the function of each subsystem and the interactions between them need to be considered early in the design phase.

2

Two-phase fluid dynamics

2.1 INTRODUCTION

Flow imaging is an example of the general area known as process tomography. It can be used to give, on a screen, an image or pattern of some property of the flow at a particular time and at some defined location in a pipe, or in a process vessel, etc. A measured property such as density, chemical composition, permittivity, etc, can be displayed by the image, and it may be possible to use the image to visualize the changing properties of the flow stream or process. This system could be extended so that a computer, analysing the 'image' data, could define a multi-phase or multi-component flow in terms of the 'flow regime'.

In this chapter we will use the description *two phase* because this most closely represents the concept of interfaces between two materials which change in time and space (terminology is discussed in Chapter 1.3).

A principal aim of this chapter is to move from the general concepts outlined above by describing the fluid dynamics of two-phase flows, i.e. flows where the phases have a large difference in density, such as gas/liquid flows and gas/solid flows. From this information the reader will appreciate the need for particular geometrical arrangements of sensors which will optimize the measurement and provide the necessary image information. Therefore we will consider the most significant flow patterns that are commonly encountered and examine the analytical or empirical relationships which enable the patterns to be predicted from the physical parameters of the flow system. From this information simple physical models of flow regimes can be constructed, which may be used for visual interpretation and for evaluating the relative performance of different sensor systems when exposed to the particular flow pattern represented by the physical model.

2.2 CATEGORIES OF FLOW IMAGES

In Chapter 1 we explained the broad approach to flow imaging so now it is appropriate to describe a few selected examples of flow regimes and imaging systems which will lead to the consideration of the detailed structure of flow images in section 2.3. In some cases we will cross the hazy boundary between flow imaging and process tomography. For example, there are close similarities between pipe flow and the flow on the shell side of a shell and tube heat exchanger, and between multi-phase flow and the separation and interfacial processes in separation tanks.

In considering the types of image that are important and the physical locations where images are required, the needs for routine process operation may differ from the needs of the research worker or process designer. For example, in a process flow where dispersed bubbles are necessary to provide a large surface area for heat or mass transfer to take place, a process operator would be interested in visualizing the flow only at the location where bubble coalescence or a change to annular flow would be likely to take place. On the other hand, the system designer would be interested in the detailed examination of this transfer from bubble to annular flow in many parts of a process in order to optimize the process design for a wide range of operating conditions. In order to illustrate and extend the above example, we will now consider a range of locations and the types of image that may be encountered.

2.2.1 Vertical pipe—cross-sectional image

In this case, the image is obtained at a defined cross-section of the pipe (figure 2.1(a)). Thus the instantaneous image function is $I(x, y)$. Since flow patterns often change with time, an additional time argument is needed, giving the image function as $I(x, y, t)$.

2.2.2 Vertical pipe—multi-sectional image

Provided that suitable sensors are available, this type of 'snapshot image' could give a three dimensional image of a length of pipe; alternatively it may be sufficient to measure the image at two planes as shown in figure 2.1(b). Images at two planes are sufficient for streamline velocities to be measured in real time (Chapter 1.9) provided that the sum of the data collecting time τ_d and the data processing time τ_p is much less than the transit time of the flow phenomenon τ_{av} between the two image planes, i.e. $\tau_d + \tau_p \ll \tau_{av}$.

2.2.3 Horizontal pipe—stratified flow

This is similar to many open channel flow situations. Where there is a well-defined separation between the high density and low density material as shown in figure 2.1(c), it is sufficient to define the measured image by the interface level h. Thus the image is defined as $I(h)$ or $I(h, t)$.

Relatively simple instrumentation may be used for this measurement. However, if the pipe fills up and higher velocities are encountered (consider,

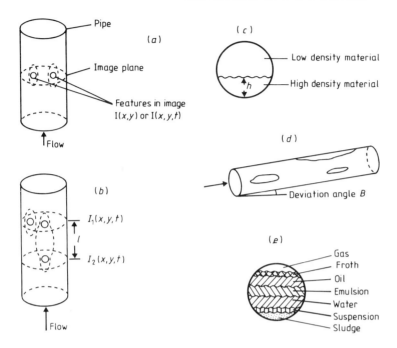

Figure 2.1 Examples of flow and process images: (*a*) vertical pipe—cross section image; (*b*) vertical pipe—multisectional image; (*c*) horizontal pipe—stratified flow; (*d*) deviated (inclined) flow; (*e*) interfaces in separation plant—gas/oil/water/solids separation.

for example, the problem of 'surcharging' which occurs when an underground sewage pipe becomes overfilled after a storm), the two phases become mixed and the more general approach to the horizontal or deviated flow situation outlined in section 2.2.4 must be considered.

2.2.4 Deviated (inclined) flow

The imaging needs of this situation (figure 2.1(*d*)) will be apparent from the examples considered above, provided that the effect of the angle of inclination is considered.   The image is thus functionally expressed by $I(x, y, \beta)$ or $I(x, y, \beta, t)$.

2.2.5 Images in separation planes

This may be considered as an example of 'process tomography' but is included here to show the close correspondence between the needs of process tomography and flow imaging. Referring to figure 2.1(*e*), we see that the vertical cross-section of a separator drum is similar to the cross section of a pipe where multiple interfaces occur. In this case, the required image information is likely to be primarily a function of the level of the interfaces.

Following the above examples we will now present a more systematic approach to flow imaging in which many parameters must be considered to determine the true nature of a multi-phase flow. Because of the high complexity of flow in a truly multi-phase situation, and because in practice the number of components in a process is frequently limited, we will consider the specific example of two-phase flow in order to derive a tractable and readily understood set of relationships.

2.3 OVERVIEW OF TWO-PHASE PHENOMENA

A description of two-phase flow should include information on the flow regime as well as information on the component fraction (the ratio of the component volume to the total volume, also referred to as void fraction). Identification of the flow regime is important in assessing the effects of two-component flow on process system performance. This is because the performance of some processes is affected not only by the relative amounts of each component, but also by the distribution of the separate phases within the flow. Examples include the large variation in oil/gas separator performance caused by slugging in the feedpipes and the superior performance of a heat exchanger pipe when exposed to an annular liquid flow than when partially exposed to the gaseous component in a two-phase flow.

A major aim of this chapter is therefore to show the links between the traditional methods of categorising two-phase flow and the newer concepts involved in flow imaging. These links are illustrated in figure 2.2. The flow process can be considered from either a microscale or a macroscale point of view. Microscale aspects include details such as bubble diameter, trajectory, and wavelength and amplitude of oscillatory motion. Macroscale aspects include broader aspects such as the flow regime (i.e. annular flow, slug flow or bubble flow), the field properties such as the symmetry of the flow that may exist in a long vertical pipe and the asymmetrical flow which may occur in many situations. The generalized field properties, such as the need to measure in one dimension (e.g. stratified flow figure 2.1(c)), two dimensions (figure 2.1(a)), or even three dimensions where the flow is very disturbed, are considered macroscale phenomena.

The conceptual interface between the flow process and the instrumentation system signifies that a clear knowledge of the flow process and the required measurements is necessary before the instrumentation system for flow imaging can be designed. The instrumentation system design portion of figure 2.2 includes information which is nearly always required for the design of any measuring system, such as accuracy, sampling time, and bandwidth. The specific needs of a flow imaging system such as the number of pixels in the image are also

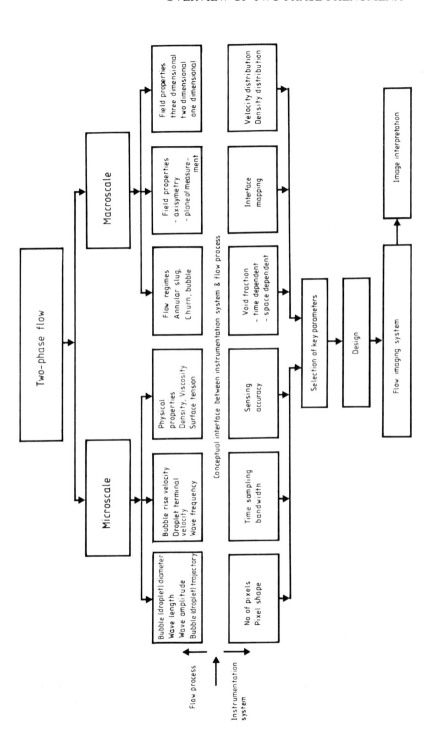

Figure 2.2 Two-phase flow image.

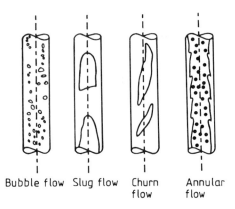

Bubble flow Slug flow   Churn   Annular
                         flow   flow

Figure 2.3   Flow regimes in vertical channel—gas/liquid.

listed. In addition, it includes the broader needs for void fraction measurement and the possible requirement to map interfaces in a flow system.

The geometrical distribution of the component interfaces is probably one of the most important aspects of two-phase flow in open channels, mass and heat exchange processes, pneumatic conveyors, fluidized bed reactors etc. The form of the interface can vary widely depending upon the flow rate, physical properties of the phases, and the geometry and inclination of the 'low channel. We will now move on to consider the nature of such macroscale flow phenomena and then we will examine quantitatively the process parameters which influence the transition between the various flow regimes.

2.4   MACROSCALE FLOW PHENOMENA

2.4.1   Regimes for upward gas/liquid flow in a vertical pipe

For upward flow in vertical pipes, four main patterns may be distinguished, see for example Hewitt and Hall-Taylor (1970). These are illustrated in figure 2.3 and their main features are discussed below.

Bubble flow. In bubble flow the gas component flows as discrete bubbles in a liquid continuum. The bubbles are usually distorted spheres.

Slug flow. When the bubble concentration in bubble flow becomes high, bubble coalescence occurs and the largest bubbles are of the same order of size as the tube diameter. Further coalescence results in the deformation of the bubble into pockets of gas which is characteristic of slug flow. Slug flow then consists of these pockets of gas attached to the wall, or bullet-shaped Taylor bubbles, separated by regions of bubble flow.

Churn flow. This is a highly disordered flow regime in which the vertical motion of the liquid is oscillatory. It possesses some of the characteristics of slug flow, the main differences being

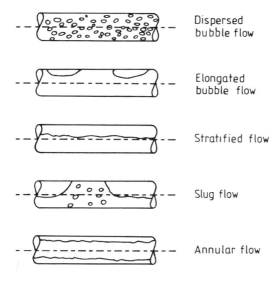

Dispersed
bubble flow

Elongated
bubble flow

Stratified flow

Slug flow

Annular flow

Figure 2.4   Flow regimes in horizontal channel—gas/liquid.

(a) The gas slugs become narrower and more irregular.
(b) The continuity of the liquid in the slug is repeatedly destroyed by regions of higher gas concentration.
(c) The thin falling film of liquid surrounding the gas slug can no longer be observed.

Annular flow. In annular flow the gas flows along the centre of the pipe. The liquid flows partially as a film along the walls of the pipe, and partially as droplets in the central gas core.

2.4.2   Regimes for co-current gas/liquid flow in a horizontal pipe

For the case of co-current flow in horizontal pipes, four flow regimes can be distinguished, see for example Annunziato and Girrardi (1987). These regimes are illustrated in figure 2.4, where slug and elongated bubble flows can be jointly classified as 'intermittent'. The main features are described below.

Dispersed bubble flow. Many spherical gas bubbles flow in the whole cross-section of the channel or in the upper zone.

Intermittent flow. This flow pattern contains both elongated bubbles (or plugs) and slugs of liquid flow regimes. According to Rudder and Hanratty (1990) these regimes are shown in figure 2.5.

Stratified flow. The gas phase flows in the upper section of the channel and the liquid in the lower section. This type of flow is generally divided into 'stratified

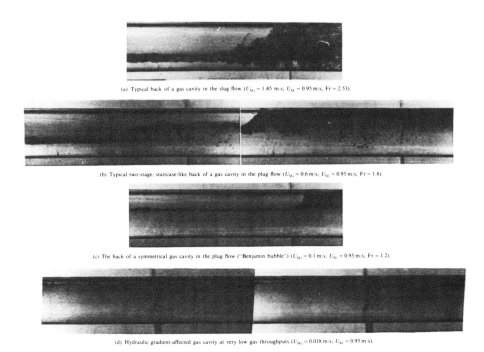

(a) Typical back of a gas cavity in the slug flow (U_{SG} = 1.45 m/s; U_{SL} = 0.95 m/s; Fr = 2.53).

(b) Typical two-stage, staircase-like back of a gas cavity in the plug flow (U_{SG} = 0.6 m/s; U_{SL} = 0.95 m/s; Fr = 1.8).

(c) The back of a symmetrical gas cavity in the plug flow ("Benjamin bubble") (U_{SG} = 0.1 m/s; U_{SL} = 0.95 m/s; Fr = 1.2).

(d) Hydraulic gradient-affected gas cavity at very low gas throughputs (U_{SG} = 0.018 m/s; U_{SL} = 0.95 m/s).

Figure 2.5   Photographs of various intermittent flow regimes within a horizontal pipe (Runner and Hawratty 1990).

smooth', where the liquid surface is smooth and without waves, and 'stratified wavy', when waves appear.

Annular flow. The liquid acts as a liquid film on the wall of the channel ('pure annular'). At high superficial liquid velocity, the film in the lower region is thicker and often assumes a wavy shape ('annular wavy').

2.4.3   Regimes for co-current gas/solids flow in a horizontal pipe

According to Tsuji and Morikawa (1982) five main flow regimes may be distinguished. These are illustrated in figure 2.6, and their main features are described below.

Dispersed flow.   At higher air velocity all particles are suspended in the dispersive phase.

Moving cluster flow.   As the velocity decreases, the pattern changes. The particles form clusters sliding on the bottom of the channel at nearly constant

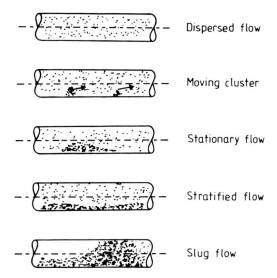

Figure 2.6 Flow regimes in horizontal channel—gas/solids system.

intervals. The velocities of the clusters are less than half the mean air velocity. Between the clusters, particles travel at higher velocity than the cluster.

Stationary cluster. As the air velocity decreases further, the particle clusters become larger and larger and almost stop sliding. The space between the large clusters and the upper wall is so narrow that it forms a kind of throat at which particles are blown off. These particles are deposited downstream to form other stationary clusters.

Stratified flow. As the air velocity decreases further, a stationary layer of deposited particles is formed on the bottom of the channel above which other particles are conveyed as a dispersion.

Slug flow. At a very low air velocity a slug of particles is formed. In the case of small particles the slug may be long enough to lead to immediate flow blockage. In the case of large particles and when the flow rate is not excessively high, the slugs can move along the pipe to sustain flow without blockage.

2.4.4   Flow regimes in fluidized beds

A fluidized bed can be regarded as a special case of two-component flow, where the solid component is quasi-stationary and the gas is the mobile component.

When gas or liquid passes upward through a bed of solid particles at a very low rate, a porous plate or grid is required to support part of the weight of the solids. These particles are stationary and form a packed bed. If the flow rate is increased, a point is reached where the solid particles are supported

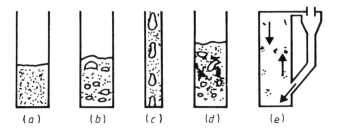

Figure 2.7 Flow regimes in fluidized bed: (*a*) particulate fluidization; (*b*) bubbling fluidization; (*c*) slugging regime; (*d*) turbulent regime; (*e*) fast fluidization.

or very nearly supported by the drag exerted by the fluid. The particles are then mobile. This is said to be the point of minimum or incipient fluidization. Beyond minimum fluidization, according to Grace (1973), there are at least five different fluidization regimes as illustrated in figure 2.7. The last four regimes are often collectively referred to as aggregative fluidization.

Not all five flow regimes will occur in every case. This is because the transition points between regimes depend not only on the gas/solid system but on other features. Grace (1973) showed that the flow regime depends mainly on the gas flow velocity and on particle size and density.

2.4.5   Regimes for shell and tube heat exchanger

A heat exchanger is one example of a process where the performance is significantly affected by the flow regime. The two-component flow in a heat exchanger is often visualized in terms of an idealized cross flow over a tube bundle. Flow patterns in cross flow have been investigated and defined by Grant (1975), as illustrated in figure 2.8. It should be emphasised that flows in practical heat exchangers differ from the idealized cross flows shown in figure 2.8(*a,b,d,e*), due to a number of effects including separation effects and fluid leaks through the spaces between the tubes and baffles. These phenomena are discussed by Hewitt (1982).

2.4.6   Flow maps

To predict the different flow regimes a flow map is used which is mainly based on visual identification of phase distribution. This is a presentation of the different flow patterns in two-dimensional space. For a gas/liquid system, commonly the superficial velocities of both phases or the momentum flux of both phases (in dimensional or non-dimensional form) are taken as co-ordinates. For the particular cases described above, flow maps are given in figures 2.9–2.13. It should be stressed that, from the physical point of view, the major factors determining the flow pattern are the velocities of the two phases.

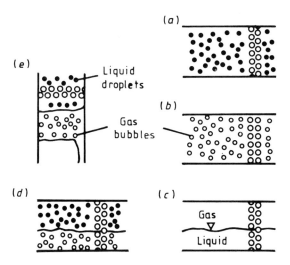

Figure 2.8 Flow regimes in cross flow: (*a*) spray; (*b*) bubbly flow, vertical and horizontal flow; (*c*) stratified flow; (*d*) stratified flow, spray flow, (*e*) slugging flow, vertical flow.

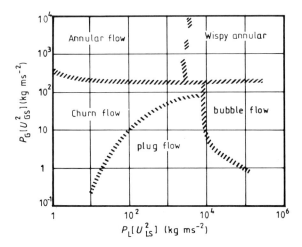

Figure 2.9 Flow-regime map obtained by Hewitt and Roberts (1969) for vertical two-phase upward flow.

Recently, a new method to recognize various flow regimes was proposed by Franca *et al* (1991). This method is based on fractal techniques and various fractal dimensions seem to be promising, for objectively discriminating between separated and intermittent flows.

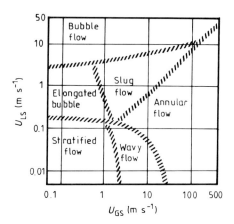

Figure 2.10 Flow pattern map obtained by Taitel and Dukler (1976) for horizontal co-current flow.

2.5 METHOD FOR PREDICTING FLOW REGIME TRANSITION FOR GAS/LIQUID SYSTEMS

In the previous section (2.4) we categorized flow regimes according to system parameters, using results obtained empirically from a limited number of experiments. It is now appropriate to take a more analytical approach so that the range of microscale and macroscale flow phenomena can be predicted at the process design stage. In this way, the value of a flow imaging system may be assessed and the range of operating conditions which it is likely to encounter may be predicted.

Based on the work of Barnea *et al* (1982a, b, 1985), Husain and Weisman (1987), Kadambi (1982), Lin and Hanratty (1986), Mishima and Ishii (1984), McQuillan and Whalley (1985), and Taitel *et al* (1980), we will now describe a method to predict the steady-state condition boundaries for a wide range of flow conditions and tube inclinations. Transition mechanisms for each individual range boundary will be presented. The transition criteria will be given for each transition in the form either of an algebraic equation, or where the equations are too complex for direct calculation, by dimensionless maps. These equations or maps incorporate the effect of flow rates, fluid properties, tube size and angle of inclination.

2.5.1 The transition from dispersed bubble flow

Dispersed bubble flow consists of well separated bubbles and usually appears at very high liquid flow rates. There are, however, conditions where small discrete bubbles also appear at low liquid rates. The gravity force plays a fundamental role in determining the flow regime. The dominant role of gravity becomes clear

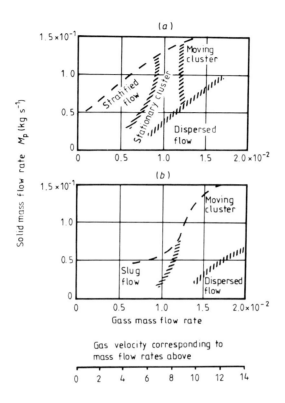

Figure 2.11 Flow regime map obtained by Tsuji and Morikawa (1982) for horizontal gas–solid co-current flow: (a) small particles $d_p \sim 0.2$ mm; (b) large particles $d_p \sim 1.5$ mm.

because changing the inclination of a pipe carrying gas and liquid can change the flow pattern from stratified to slugging.

Dispersed bubble flow at high liquid flow rates is observed over the whole range of pipe inclination, while at low liquid flow rates bubble flow regimes are observed only in vertical and near-vertical flows in relatively large diameter pipes. The bubble flow regime can exist in these pipes provided two conditions are met:

(i) The Taylor bubble (section 2.4.1) velocity exceeds the bubble velocity. For this condition coalescence of the small bubbles around the nose of Taylor

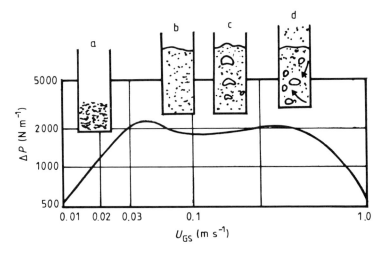

Figure 2.12   Flow regime map obtained by Kuni and Levenspeil for fluidized sand: (*a*) fixed bed; (*b*) particulate fluidization; (*c*) bubbling fluidization; (*d*) turbulent regime.

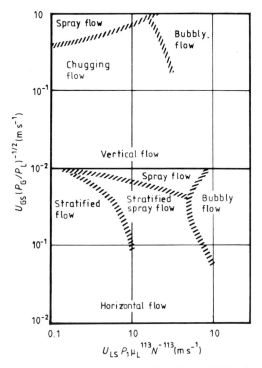

Figure 2.13   Flow regime maps obtained by Grant (1975) for cross flow.

bubbles does not take place. This condition is satisfied in large diameter (*D*) pipes:

$$D \geqslant 19 \left[\frac{(\rho_L - \rho_G \sigma)}{\rho_L^2 g} \right]^{1/2} \tag{2.1}$$

where D is the pipe diameter, ρ_L and ρ_G are the liquid and gas densities, and σ is the surface tension.

(ii) The angle of inclination (from the horizontal) is large enough to prevent migration of the bubbles to the top wall of the pipe. The critical angle, β_c, can be calculated from:

$$\frac{\cos \beta_c}{\sin^2 \beta_c} = \frac{3}{4} \left[\cos 45° \right] \left[\frac{U_T^2}{g} \right] \left[\frac{C_L \gamma}{d} \right] \tag{2.2}$$

where d is the bubble diameter, U_T is the bubble velocity, C_L is the 'lift' coefficient, and γ is the distortion coefficient of the bubble, U_T is given by the relation:

$$U_T = 1.53 \left[\frac{g(\rho_L - \rho_G)\sigma}{\rho_L^2} \right]^{1/4}. \tag{2.3}$$

The average value suggested for C_L is 0.8 and γ ranges from 1.1 to 1.5, based on observation by Streeter (1961).

It was shown by Taitel *et al* (1980) that transition from bubbly flow to slug flow takes place when the gas void fraction exceeds a critical value of $\alpha = 0.25$. This transition is given by:

$$U_{LS} = \left(\frac{1-\alpha}{\alpha} \right) U_{GS} - 1.53(1-\alpha) \left[\frac{g(\rho_L - \rho_G \sigma)}{\rho_L^2} \right]^{1/4} \sin \beta \tag{2.4}$$

where the angle β is positive for upward and negative for downward flow, U_{LS} and U_{GS} are the liquid and gas superficial velocities defined as follows:

$$U_{LS} = U_L(1-\alpha) \qquad U_{GS} = U_G \alpha.$$

At high liquid flow rates dispersed bubbles exist even for $\alpha > 0.25$, due to turbulence which causes bubble break-up and prevents agglomeration. The unified model for this transition was recently proposed by Barnea (1987). The result is:

$$d_C \geqslant \left[0.725 + 4.15 \left(\frac{U_{GS}}{U_M} \right)^{1/2} \right] \left(\frac{\sigma}{\rho_L} \right)^{3/5} \left(\frac{2f_M}{U_M^3} \right)^{-2/5} \tag{2.5}$$

where d_C is the diameter of the dispersed bubble, U_M is the mixture velocity and f_M is the friction factor based on the mixture velocity.

The bubble diameter on the transition boundary, d_C, is a function of the liquid velocity and the angle of inclination, the value of d_C is taken as the smaller of

d_{CD} and d_{CB} where d_{CD} is the critical bubble size above which the bubble is deformed:

$$d_{CD} = 2\left[\frac{0.4\sigma}{(\rho_L - \rho_G)g}\right]^{1/2} \tag{2.6}$$

and d_{CB} is the critical bubble size below which migration of bubbles to the upper part of the pipe is prevented:

$$d_{CB} = \frac{3}{8}\left[\frac{\rho_L}{\rho_L - \rho_G}\right]\frac{f_M U_M^2}{g\cos\beta}. \tag{2.7}$$

The transition boundary (equation (2.5)) is valid for $0 < \alpha < 0.52$. At the upper limit the maximum volumetric packing density of the bubbles is reached and coalescence occurs at high turbulence levels. The transition curve that characterizes this condition is:

$$U_{LG} = \frac{1 - \alpha}{\alpha}U_{GS} \tag{2.8}$$

where $\alpha = 0.52$.

2.5.2   The stratified/non-stratified transition

For horizontal and slightly inclined pipes, Taitel and Dukler (1976) suggested that the transition from equilibrium stratified flow is due to Kelvin–Helmholtz instability. They suggested that the criterion at which the wave will grow, so that the transition from stratified flow occurs is:

$$F^2\left[\frac{1}{(1 - \tilde{h}_L)^2}\tilde{U}_G^2\frac{\left(\frac{d\tilde{A}_L}{d\tilde{h}_L}\right)}{\tilde{A}_G}\right] \geqslant 1 \tag{2.9}$$

with

$$\tilde{A}_L = \frac{A_L}{A} \qquad \tilde{A}_G = \frac{A_G}{A} \qquad \tilde{h}_L = \frac{h_L}{h}$$

where A_L, A_G represents a channel cross section occupied by liquid or gas respectively. A is a channel cross section, h_L is the thickness of the liquid layer, h is the width of the channel, F is the Froude number modified by the density ratio

$$F = \left(\frac{\rho_G}{\rho_L - \rho_G}\right)^{1/2}\frac{U_{GS}}{(Dg\cos\beta)^{1/2}}. \tag{2.10}$$

Note that all terms in the square brackets in equation (2.9) are functions of $\tilde{h}_L$ only. Once the gas and liquid flow rates, fluid properties, inclination angle, and pipe size are specified, the equilibrium liquid level is found by solving the

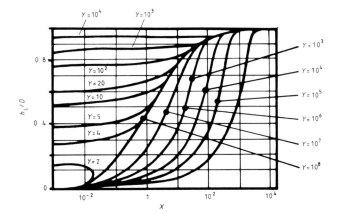

Figure 2.14 Equilibrium level in stratified flow.

momentum equations for each component in stratified flow. This equilibrium level is given in terms of X and Y:

$$X^2 = \frac{\frac{4}{D} f_{LS} \left(\frac{\rho_L U_{LS}^2}{2} \right)}{\frac{4}{D} f_{GS} \left(\frac{\rho_G U_{GS}^2}{2} \right)} = \frac{\left(\frac{dp}{dx} \right)_{LS}}{\left(\frac{dp}{dx} \right)_{GS}} \qquad (2.11a)$$

and

$$Y = \frac{(\rho_L - \rho_G) g \sin \beta}{\left(\frac{dp}{dx} \right)_{GS}} \qquad (2.11b)$$

$(dp/dx)_S$ designates the pressure drop of one component flowing in the pipe, and f_{LS} and f_{GS} are the friction factor coefficients for single component liquid or gas flows respectively. This can be calculated by using equivalent correlations.

The equilibrium level $\tilde{h}_L$ as a function of X is shown in figure 2.14 for various values of the parameter Y. The transition criterion given by equation (2.9) is presented in figure 2.15 by a curve A on a two-dimensionless map with f and h_L as co-ordinates.

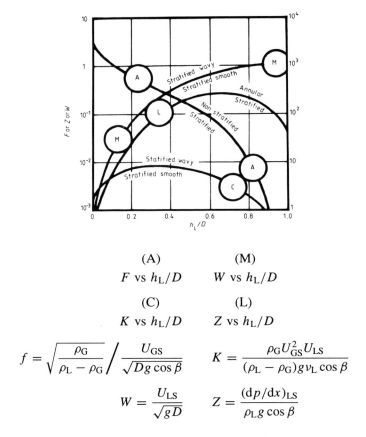

(A)   (M)

F vs h_L/D W vs h_L/D

(C)   (L)

K vs h_L/D Z vs h_L/D

$$f = \sqrt{\frac{\rho_G}{\rho_L - \rho_G}} \Bigg/ \frac{U_{GS}}{\sqrt{Dg\cos\beta}} \qquad K = \frac{\rho_G U_{GS}^2 U_{LS}}{(\rho_L - \rho_G)g\nu_L\cos\beta}$$

$$W = \frac{U_{LS}}{\sqrt{gD}} \qquad\qquad Z = \frac{(\mathrm{d}p/\mathrm{d}x)_{LS}}{\rho_L g\cos\beta}$$

Figure 2.15   Generalized transition bounaries (after Barnea 1987).

2.5.3   The stratified-annular transition

Transition mechanism A in figure 2.15 presents a criterion under which finite waves on stratified liquid flow are expected to grow resulting in either slug or annular flow. Once the physical properties, pipe diameter, and inclination angle are specified, the predicted transition line can also be plotted on the U_{LS} and U_{GS} map (figure 2.10).

The region below curve A is normally a stratified flow. However, at steep downward inclination, another mechanism comes into play, by which stable stratified flow is seen to change into annular flow at relatively low gas flow rates. This latter transition boundary is applicable only within the region bounded by curve A.

At a constant flow rate, as the downward inclination angle is increased, the stratified liquid thickness becomes small and the liquid velocity (U_L) high.

Under these conditions droplets are torn from the wavy turbulent interface and may be deposited on the upper wall, resulting in an annular film. The condition for this type of annular flow to take place is:

$$U_L^2 > \frac{[gD(1 - \tilde{h}_L) \cos \beta]}{f_L} \tag{2.12}$$

or in dimensionless form:

$$Z = \frac{\left(\frac{dp}{dx}\right)_{LS}}{\rho_L g \cos \beta} > 2 \left(\frac{\tilde{A}_L}{\tilde{A}}\right)(1 - \tilde{h}_L)\left(\frac{f_{LS}}{f_L}\right) \tag{2.13}$$

where f_L is the liquid fraction factor coefficient and f is the coefficient that refers to single-component liquid flow in the pipe.

All quantities on the right hand side of equation (2.13) depend only on $\tilde{h}_L$. The curve describing the relation between Z and K which satisfies equation (2.13) is designated as boundary L in figure 2.15. In the region below curve L stratified flow exists. Transition L is shown to appear only at steep downward inclinations as stratified flow is changed gradually to annular flow. For the limiting case of vertical downward flow, $Z \to \infty$ and stratified flow disappears completely. For small angles of downward inclination and for upward inclination, transition L is outside the region bounded by transition A and therefore is not applicable. These transitions L can be applied in the whole range of pipe inclinations.

2.5.4   Transition from annular to intermittent flow

This transition is assumed to occur when the gas core is blocked at any location by the liquid. Blockage of the gas core may result from two possible mechanisms
(a) Instability of the liquid film, due to partial downflow of liquid near the wall causing blockage at the entrance.
(b) Blockage of the gas core as a result of a large supply of liquid in the film.

Referring to figure 2.16 the condition for the instability of the liquid film in annular flow (mechanism A) has the form

$$Y = \frac{1 + 75\alpha_L}{(1 - \alpha_L)^{5/2\alpha_L}} \frac{1}{\alpha_L^3} X^2 \tag{2.14}$$

and

$$Y \geqslant \frac{2 - \frac{3}{2}\alpha_L}{1 - \frac{3}{2}\alpha_L} X^2. \tag{2.15}$$

The value of α_L that satisfies the condition for the film instability is obtained from equation (2.15). Equation (2.14) gives the steady-state solution for the

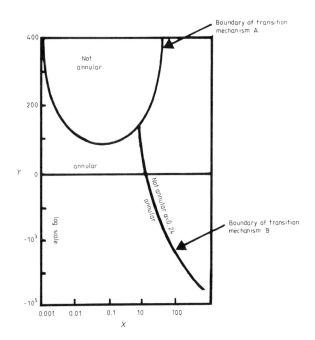

Figure 2.16   Generalized map for annular-intermittent transition.

liquid hold-up in the annular flow. Slugging, that is blockage of the gas core by liquid lumps (mechanism B), will occur when the liquid supply in the film is large enough to provide the liquid needed to bridge the pipe. The condition for slugging has the form:

$$\frac{A_L}{A R_{sm}} = \frac{\alpha_L}{R_{sm}} > 0.5 \tag{2.16}$$

where R_{sm} is the minimum liquid holdup within the formed liquid slug that will allow complete bridging of the gas passage. This minimum value is related to the maximum bubble volume in the liquid slug and equals approximately 0.48. Lower values of R_{sm} will make slugging impossible due to the high void fraction. The transition boundaries are shown in figure 2.16 in terms of dimensionless co-ordinates X and Y (transition mechanism 'A'). Equation (2.14) with the value of $\alpha_L = 0.58 R_{SM}$ yields the condition where the blockage occurs (mechanism 'B'). Equation (2.14) is plotted for a constant value $\alpha_L = 0.24$ within the stable zone (the 'B' line).

2.5.5   Subregions within intermittent flow

The intermittent pattern is usually subdivided into elongated bubble, slug, and churn flow. Basically, these three flow patterns have the same configuration with respect to the distribution of the gas and liquid interfaces. In these flow patterns, slugs of liquid are separated by large bullet-shaped bubbles. In slug flow the liquid bridges contain small gas bubbles. The 'elongated bubble' pattern is considered as the limiting case of slug flow, when the liquid slug is free of entrained bubbles, while churn flow takes place when the gas void fraction within the liquid slug reaches a maximum value above which occasional collapse of the liquid slug occurs. It was suggested by Barnea and Brauner (1985) that the gas holdup on the transition line from dispersed bubbles is the maximum holdup that the liquid can accommodate as fully dispersed bubbles at a given mixture velocity $U_M = U_{GS} + U_{LS}$. The curve of constant U within the intermittent region represents the locus where the gas holdup α_S is constant and equal to the holdup of the dispersed bubble pattern at the transition boundary. Once the fluid properties and pipe size are set, α_S can be obtained from equation (2.5) to yield:

$$\alpha_S = 1 - R_{sm} = 0.058 \left[d_C \left(2 f_M \frac{U_M^2}{D} \right)^{2/5} \left(\frac{\rho_L}{\sigma} \right)^{3/5} - 0.725 \right]^2 . \qquad (2.17)$$

This equation predicts the equilibrium gas void fraction within the main body of the liquid slug. When the elongated bubble-slug reaches the maximum bubble volumetric packing ($\alpha = 0.52$), the continuity of the very aerated liquid slug is destroyed by bubble agglomeration and the formation of regions of high gas concentration within the liquid slug, thus resulting in transition to churn flow.

2.5.6 Subregions in stratified flow

Subregions in stratified flow are usually defined as stratified smooth and stratified wavy. Waves may be formed on a smooth liquid interface due to the action of gas windage (typical of horizontal pipes) or as a result of fast liquid flow (typical of downward inclined pipes). It was shown by Taitel and Dukler (1976) that the condition for wave generation by windage is:

$$U_G \geqslant \left[4 \frac{\rho_L - \rho_G}{\rho_G} \frac{g \cos \beta}{s} \frac{\mu_L}{U_L} \right]^{1/2} \qquad (2.18)$$

or in dimensionless form:

$$K \geqslant \frac{2}{\tilde{U}_G \tilde{U}_L^{1/2} s^{1/2}} \qquad (2.19)$$

where μ_L is the liquid viscosity and s is a sheltering coefficient.

K is the product of the modified Froude number and the square root of the superficial Reynolds number of the liquid:

$$K^2 = F^2 Re_{LS} = \frac{\rho_G}{\rho_L - \rho_G} \cdot \frac{U_{GS}^2}{Dg \cos \beta} \cdot \frac{dU_{LS}}{\mu_L} . \qquad (2.20)$$

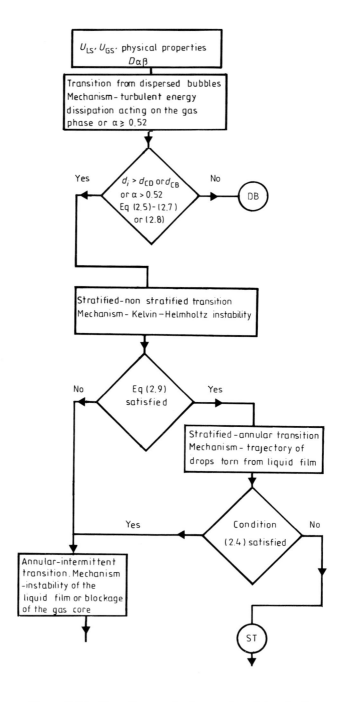

Figure 2.17 Flow diagram for flow regime determination.

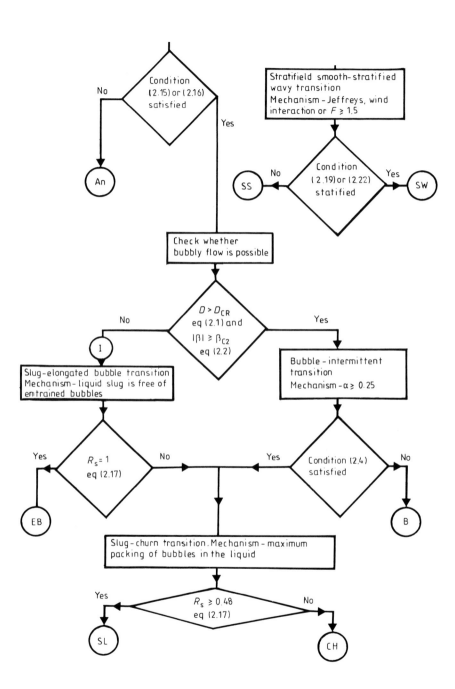

Figure 2.17   Continued.

The transition between stratified smooth and stratified wavy, resulting from the 'wind' effects, is mapped on a K versus h co-ordinate system and is shown in figure 2.15 as curve C.

With gas flow or with turbulent flow in smooth pipes, waves can be developed on a liquid in a sloping pipe even in the absence of interfacial shear. Barnea *et al* (1982*a*) adopted the following criteria for wave inception:

$$F = \frac{U_L}{(g\tilde{h}_L)^{1/2}} \geqslant 1.5 \tag{2.21}$$

or

$$W = \frac{U_{LS}}{(gD)^{1/2}} \geqslant 1.5\tilde{h}_L^{1/2}\frac{\tilde{A}_L}{\tilde{A}} \tag{2.22}$$

The right-hand side of equation (2.22) depends only on $\tilde{h}_L$, thus the locus of W versus $\tilde{h}_L$ which satisfies equation (2.22) is the smooth/stratified boundary and is shown as curve M in figure 2.15.

2.5.7   Summary of methods to predict the flow pattern

The rather complex set of procedures outlined in sections 2.5.1–2.5.6 above is summarized in the form of a flow diagram in figure 2.17. This should assist the reader in determining the range of conditions likely to be encountered in any system.

2.6   MICROSCALE STRUCTURE

2.6.1   Drops and bubbles

Drops and bubbles rising, or falling freely under gravity, in a Newtonian liquid, are commonly considered to belong to one of three broad shape regimes: spherical, ellipsoidal or spherical cap. In reality, these regimes cover shapes that are not strictly spheres, ellipsoidal or spherical caps. To be included within the spherical regime, the drop or bubble must be rounded with a minimum aspect (height to width) ratio of 0.9. The ellipsoidal regime includes flattened drops and bubbles with a concave surface (viewed from inside) around the entire periphery. The shapes may, however, lack fore-and-aft symmetry, and oscillations, dilations, or wobbling may occur. It was shown by Grace (1973) that a convenient mapping of these three principal regimes, as well as of certain subregions, can be obtained by plotting the drop or bubble Reynolds number:

$$\text{Re} = \frac{\rho d_e U_T}{\mu} \tag{2.23}$$

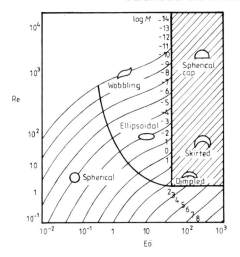

Figure 2.18 Shape regimes and subregions for liquid drops and gas bubbles rising falling freely through liquids under gravity by Clift *et al* (1976).

versus the Eötvos number:

$$\text{Eö} = \frac{g\Delta\rho d_e^2}{\sigma} \tag{2.24}$$

with a fluid group property:

$$M = \frac{g\mu^4 \Delta\rho}{\rho^2 \sigma^3} \tag{2.25}$$

as a parameter. Here, U_T is the terminal rising or falling velocity of the drop or bubble, de the sphere volume-equivalent diameter ($d_e = [6V_p/\pi]^{1/3}$), and $\Delta\rho = |\rho - \rho_p|$, where V_p is the bubble volume. Unsubscripted properties refer to the outer liquid phase, while the subscript p is used to denote the bubble or drop (dispersed phase).

The regime map given by Grace *et al* (1976) appears in figure 2.18. The heavy lines delineate the three principal shape regimes. Subregions are shown within the ellipsoidal and spherical cap regimes corresponding to wobbling conditions, skirt and ellipsoidal caps.

For a spherical bubble or drop (of small diameter) its terminal velocity (or settling velocity) is derived as:

$$U_T = \frac{2}{3} \frac{ga^2 \Delta\rho}{\mu} \frac{k+1}{3k+2} \tag{2.26}$$

where k = viscosity ratio, μ_p/μ. Experimental results showing the terminal velocity of intermediate size area bubbles in water rising freely under gravity are shown in figure 2.19. The ellipsoidal regime for this commonly encountered system covers bubble volume-equivalent diameters ranging from about 1–17 mm. The onset of deformation with increasing drop or bubble volume is due to

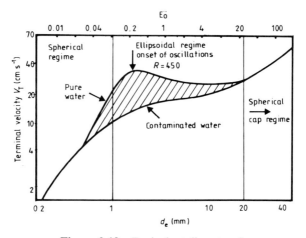

Figure 2.19 Equivalent diameter d_e.

inertia forces. There are many data in the literature pertaining to terminal rising or settling velocities of drops and bubbles in liquid that apply to the ellipsoidal regime. Grace (1973) and Clift *et al* (1978) presented a method of predicting terminal velocities. In this method wall effects are not included. To obtain the terminal velocities in the ellipsoidal regime, one first calculates a dimensionless group:

$$H = \frac{4}{3}\text{Eö}\,M^{-0.149}\frac{\mu}{\mu_W} \tag{2.27}$$

where $\mu_W = 9 \times 10^{-6}\,\text{kg}\,\text{ms}^{-1}$ is the viscosity of water.

The terminal velocity is then given by:

$$U_T = \frac{\mu}{\rho d_e}M^{-0.149}(J - 0.857) \tag{2.28}$$

where

$$J = 0.94 \times H^{0.757} \qquad (2 < H < 59.3) \tag{2.29}$$

or

$$J = 3.42 \times h^{0.441} \qquad (H > 59.3). \tag{2.30}$$

The above method is recommended for systems with $M < 10$, Re > 0.1, Eö < 40. Drops and bubbles travel more rapidly in pure liquids than in contaminated liquids, as explained by Skelland and Huang (1977). This is caused by surface-active agents which form on the boundaries between the components. The terminal velocity in pure liquids, denoted by a superscript p, may be obtained from:

$$U_T^p = U_T\left(1 + \frac{\Gamma}{1 + k}\right) \tag{2.31}$$

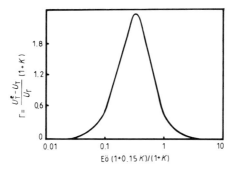

Figure 2.20 Correction factor relating terminal velocity of drops and bubbles in pure liquids to those corresponding contaminated systems by Clift *et al* (1978).

where U_T is obtained from equations (2.27) to (2.30) and Γ is given by figure 2.20.

The onset of secondary motion of drops and bubbles coincides with the onset of vortex shedding from the wake. Secondary motion may involve shape dilations (usually called oscillations), rocking from side to side, spiral motion, or some combination of these three. Thus, the paths followed may be rectilinear, zigzagging in a plane, or spiral. The natural frequency of small oscillations in the absence of viscous force is:

$$f = \left[\frac{48\sigma}{\pi^2 d_e^3 (2\rho + 3\rho_p)} \right]^{1/2}. \tag{2.32}$$

Wall effects tend to cause retardation of drops and bubbles, and some elongation in the vertical direction. For Re > 200, the terminal velocity of an ellipsoidal drop or bubble subject to wall effects as a ratio to that for a container of infinite cross-section, is given approximately by:

$$\overline{U}_T = \left[1 - \left(\frac{d_e}{D} \right)^2 \right]^{3/2} h \tag{2.33}$$

where D is the diameter of a circular container or the equivalent hydraulic diameter of a non-circular container (4 × cross-sectional area divided by perimeter). This relationship is recommended for a $d_e/D < 0.6$, Eö < 40, and Re > 200. For $d_e/D > 0.6$, the drop or bubble must be treated as a slug.

Interaction of ellipsoidal bubbles and drops is complicated by secondary motion such as zigzagging or helical rise. Hence two bubbles or drops that begin motion along a common vertical axis may not remain aligned (de Nevers and Wu 1971, Otake *et al* 1977). In addition, when collision occurs it may not be followed by coalescence.

Coalescence is governed by the rate of drainage of the thin film of continuous phase between the bubbles or drops at small separation. More details concerning the coalescence of ellipsoidal bubbles are given by Zabel *et al* (1973) and Yip *et al* (1979).

The spherical cap regime (figure 2.18) covers most bubbles and drops with Eö > 40. In general, this means volumes greater than about 3 cm^3 or d_e > 18 mm. Such large bubbles are of interest in liquified metals, fluidized beds, and underwater explosions. For spherical cap bubbles with Re > 150, the front surface is very much like a segment of a true sphere, while the rear surface is almost flat, as shown schematically in figure 2.18. The wake's angle Θ (defined as the angle between the vertical axis and the line joining the centre of curvature of an arc fitting the front portion to the outer rim of the cap) is then very nearly 50°. This angle in degrees can be predicted from an empirical relationship suggested by Clift *et al* (1978):

$$\theta = 50° + 190° \exp(-0.63 Re^{0.4}). \tag{2.34}$$

The terminal velocity of spherical bubbles obtained by Collins (1967) is:

$$U_T = 0.652 \left(\frac{g Re \Delta \rho}{\rho} \right)^{1/2}. \tag{2.35}$$

In systems where the fluid group property M defined by equation (2.25) satisfies $M > 0 \cdot 1$, the bubbles and drops may trail thin films of the dispersed phase, called skirts, as shown schematically in figure 2.18. Skirt formation occurs when viscous forces acting on the outer rim are sufficient to overcome the restraining effect of surface or interfacial tension force. For gas bubbles, skirt formation has been found to require Re > 9 and

$$W_e = \frac{\mu U_T}{\sigma} > 2.32 + \frac{11}{(Re - 9)^{0.7}} \qquad W_e = \frac{\rho U_T^2}{\sigma d_e} \tag{2.36}$$

while the corresponding conditions for liquid drops are Re > 4 and W_e/Re > 2.3.

Containing walls lead to elongation, smaller wakes, and a reduction in the terminal velocity of large bubbles, which can be predicted by Collins (1967) using the following formula:

$$\frac{U_T^w}{U_T} = 1.13 \exp \left(\frac{-d_e}{D} \right). \tag{2.37}$$

Drops falling freely under gravity through air remain very nearly spherical for Eö $= g \Delta \rho d_e^2 / \sigma$ < 0.4. Careful experimental results have been reported by Beard and Pruppacher (1969). As Eö increases above 0.4 some distortion from the spherical can be detected. Flattening occurs primarily at the leading (lower) surface so that the shape then lacks fore-and-aft symmetry. The shape can be represented by two half-spheroids sharing a common horizontal circular cross-section of a radius having semi-minor axes b_1 and b_2. The height-to-maximum width ratio is given approximately by:

$$E = \frac{b_1 + b_2}{2a} = \left[1 + 0.18(\text{Eö} - 0.4)^{0.8}\right]^{-1} \qquad 0.4 < \text{Eö} < 8. \qquad (2.38)$$

The ratio of the lower vertical semi-axis to the total height is:

$$\frac{b_1 + b_2}{2a} = \left[1 + 0.24(\text{Eö} - 0.5)^{0.8}\right]^{-1} \qquad 0.5 < \text{Eö} < 8. \qquad (2.39)$$

For Eö < 0.5 terminal velocities can be calculated as for rigid spheres. For larger drops, the following equations are recommended:

$$\text{Re} = 1.62\text{Eö}^{0.755} M^{-0.25} \qquad 0.5 < \text{Eö} < 1.84 \qquad (2.40)$$

$$\text{Re} = 1.83\text{Eö}^{0.555} M^{-0.25} \qquad 1.84 < \text{Eö} < 5.0 \qquad (2.41)$$

and

$$\text{Re} = 2.0\text{Eö}^{0.5} M^{-0.25} \qquad \text{Eö} > 5.0. \qquad (2.42)$$

In dimensionless form equation (2.42) gives:

$$U_T = 2\left(\frac{g\sigma\Delta\rho}{\rho^2}\right)^{0.25} \qquad \text{Eö} > 5.0. \qquad (2.43)$$

The terminal velocity of large drops in gases eventually becomes independent of both drop size and gas viscosity. Drops tend to become unstable and break up in stagnant media when Eö reaches or exceeds about 16.

Finally, an example of liquid droplet oscillations in a gaseous environment studied by using a multi-exposure method, Becker *et al* (1991), is shown in figure 2.21. In this picture the asymmetry of the oscillations is caused by non-linear effects.

2.6.2 Waves on the gas/liquid interface

There are several possible mechanisms for energy transfer across an interface. Based on the dominant hydrodynamic forces involved, the causes of the generation of surface waves can be classified as the inviscid flow pressure, viscous shear, inviscid stress, viscous stress, and turbulent fluctuation of pressure.

It is well known that turbulent fluctuations of pressure and inviscid stress are important for the generation of longer waves on deep water. On the other hand, for viscous fluids, wave formation is well predicted by the Kelvin–Helmholtz instability. Capillary and short gravity waves appear to be generated initially by the viscous stress, which is also important for waves over a thin liquid film.

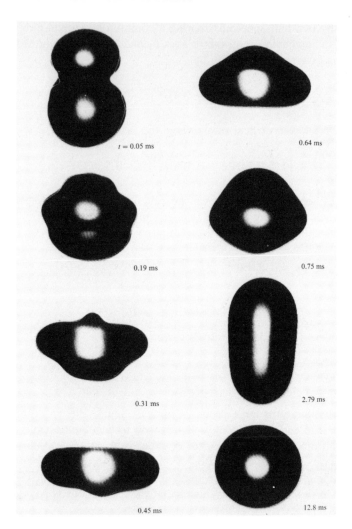

Figure 2.21 Photographs of oscillating droplets at different moments after the breakoff from the jet (Becker *et al* 1991).

In general, for waves on an initially flat interface, the surface-tension force is always stabilizing, since the flat interface has minimum surface area, and the surface-tension force acts to resist any deformation from the equilibrium configuration. The gravity term is stabilizing only if the upper fluid is lighter than the lower fluid (otherwise Rayleigh–Taylor instability can be observed). On the other hand, relative motion between the fluids is destabilizing.

A wavy liquid film (especially in annular flow) can be entrained into a gas flow in a number of different ways. Hydrodynamic and surface-tension forces

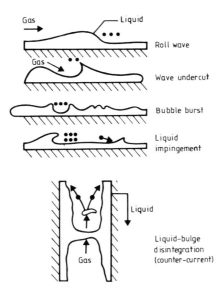

Figure 2.22   Various mechanisms of entrainment.

govern the motion and deformation of the wave crests. Under certain conditions, these forces lead to an extreme deformation of the interface, which results in a break up of a portion of a wave into several droplets. The forces acting on the wave crests depend on the flow pattern around them and on the shape of the interface. In general, the following five basic types of entrainment mechanisms, which are shown in figure 2.22, can be considered:

- shearing off of the tops of the roll waves by the gas flow,
- undercutting of the liquid film by the gas flow,
- bursting of gas bubbles,
- impingement of large drops,
- disintegration of a liquid bulge by gas flow in a counter-current situation.

A detailed entrainment-inception model was developed by Ishii and Grolmes (1975) by considering these mechanisms.

Recently Hewitt *et al* (1990) studied the structure of the gas–liquid film flow in horizontal annular flow. The results are shown in figure 2.23 where also a new conceptual picture of gas–liquid interface is shown.

2.7   MATHEMATICAL MODELLING OF TWO-PHASE FLOW

In the previous section 2.3 we saw how the macroscale structure (flow regime) could be predicted from the parameters and operating conditions of a two-phase

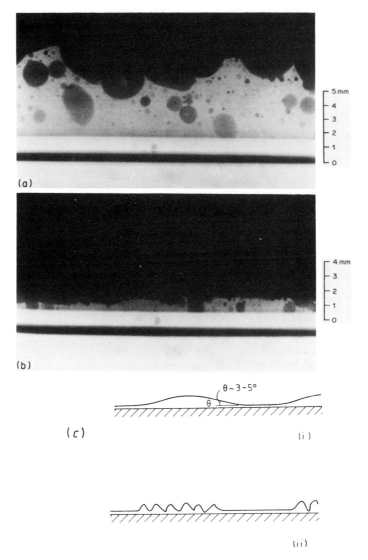

Figure 2.23 Instantaneous structure of the liquid film at the bottom part of the tube: (a) air flow rate $= 0.026\,\mathrm{kg\,s^{-1}}$ and water flow rate $= 0.100\,\mathrm{kg\,s^{-1}}$; (b) air flow rate $= 0.041\,\mathrm{kg\,s^{-1}}$ and water flow rate $= 0.100\,\mathrm{kg\,s^{-1}}$; (c) conceptual picture of the gas liquid interface: (i) traditional view ; (ii) proposed view (Hewett *et al* 1990).

flow process. A further refinement would be to predict the microscale flow structure, thus giving more detailed information of the flow structure and profile over the cross-section of a pipe. This additional information is, in principle, available from more refined mathematical models of the flow process.

Rational derivations of the instantaneous space-averaged and local time-averaged equations are given by Delhaye (1974), Drew (1983), Ishii (1975), Nigmatulin (1978) among others. The philosophy behind averaging is that the exact equations contain details of the flow which are of no use on the scale of interest. Averaging the equations gives a set of 'filtered equations' which do not contain the unwanted details of the flow. The price paid for the lack of unwanted detail in the averaged equations is that several terms appear in the averaged equations which are not determined by the averaging process. These terms contain the effects of the lost information, and must be determined from appropriate empirical equations.

In two-phase flow, an interface can be considered as a surface of discontinuity. The balance laws for each phase are expressed in terms of partial differential equations but at the interface they are formulated in terms of jump conditions.

The derivation of the local instantaneous equations starts with the integral balance laws written for a fixed control volume containing both phases. These integral laws are then transformed by means of the Leibnitz rule and the Gauss theorem to obtain a sum of two volume integrals and a surface integral. The volume integrals lead to the local instantaneous jump conditions valid on the interface only.

Jump conditions constitute a characteristic feature of two-phase flow analysis and provide relations between the phase interaction terms which appear in the averaged equations (Ishii 1975).

The choice of a 'picture' (flow image) for a two-phase flow is essentially a choice of: (i) geometric properties (e.g. axisymmetry, cylindrical interfaces in annular flow), (ii) kinematic properties such as local relative velocity between phases or (iii) thermal properties (e.g. saturation conditions for one phase or for both phases).

Categories (ii) and (iii) represent the kinematic or thermal non-equilibria between the phases. The selection of a given kinematic and/or thermal pattern is equivalent to a choice of solutions for the mathematical model.

2.7.1   The averaging process

As mentioned above, in order to obtain equations which do not contain the full details of the flow, it has become customary to apply some sort of averaging method. The generic averaging method according to Drew (1983) is presented below.

Let $\langle\ \rangle$ denote an averaging process so that if $f(x, t)$ is an exact microscopic field, then $\langle f \rangle(x, t)$ is the corresponding averaged field. An averaging process assigns average values to certain variables. The ensemble, or set of possible outcomes, can be taken to be the possible flows in some apparatus where the prescribed initial and boundary conditions are equivalent in some sense. For example, for spherical particles it may be necessary to give the statistical distribution of the positions and velocities of the centres of the particles at time

$t = 0$, such that the average number density and average particle velocity is the same for all equivalent flows. We shall assume that there is some ensemble, with some appropriate weighting $w(\omega)d\omega$, the average of f is given by:

$$\langle f \rangle (\boldsymbol{x}, t) = \int_{\Omega} f(\boldsymbol{x}, t, \omega)\, w(\omega)\, d\omega. \tag{2.44}$$

Assuming that a flow is nearly steady, so that a time translation makes no essential difference in the ensemble averaging process, we assign a weight $w(\tau)$ to the likelihood that the outcome at time τ is $f(\boldsymbol{x}, t - \tau)$, where $f(\boldsymbol{x}, t)$ is the outcome at time tin the general case. The average of f is then taken to be:

$$\langle f \rangle_t (\boldsymbol{x}, t) = \int_{-\infty}^{\infty} f(\boldsymbol{x}, t - \tau)\, w(\tau)\, d\tau. \tag{2.45}$$

This is a classical time averaging; it is often used with:

$$w(\tau) = \begin{cases} \frac{1}{T} & \text{if } 0 \leqslant \tau \leqslant T \\ 0 & \text{otherwise} \end{cases} \tag{2.46}$$

although other averages are possible.

If there are no boundaries in the flow (i.e. boundary effects are unimportant), then small spatial translations should make no difference in the ensemble. An analogy to the above, the average:

$$\langle F \rangle_S (\boldsymbol{x}, t) = \int_{R^3} f(\boldsymbol{x} + \boldsymbol{s}, t)\, w(\boldsymbol{s})\, d\boldsymbol{s} \tag{2.47}$$

can be defined. This is the classical space average; it is often used with:

$$w(\boldsymbol{s}) = \begin{cases} \frac{1}{V} & \text{if } \boldsymbol{s} \in V \\ 0 & \text{otherwise} \end{cases} \tag{2.48}$$

where V is some volume (for example, a sphere).

Again, other averages are possible. Thus, in some sense, ensemble averaging contains space and time averages as special cases. The averaging process is assumed to satisfy:

$$\begin{aligned} \langle f + g \rangle &= \langle f \rangle + \langle g \rangle \\ \langle \langle f \rangle g \rangle &= \langle f \rangle \langle g \rangle \\ \langle c \rangle &= c \\ \langle \frac{\partial f}{\partial t} \rangle &= \frac{\partial \langle f \rangle}{\partial t} \\ \langle \frac{\partial f}{\partial x} \rangle &= \frac{\partial \langle f \rangle}{\partial}. \end{aligned} \tag{2.49}$$

The first three of these relations are called Reynold's rules, the fourth is called Leibnitz's rule, and the fifth is called Gauss's rule.

In order to apply the average to the equations of motion for each phase, we introduce the phase function $X_k(x, t)$ which is defined to be:

$$X_k(x, t) = \begin{cases} 1 & \text{if } x \text{ is in phase } k \text{ at time } t \\ 0 & \text{otherwise} \end{cases}.$$
(2.50)

We shall deal with X_k as a generalized function, in particular in regard to differentiation. The derivative of a general function can be defined in terms of a set of 'test functions', which are 'sufficiently' smooth and have compact support. Then $\partial X_k/\partial t$ and $\partial X_k/\partial x$ are defined by:

$$\int_{3R*R} \frac{\partial X_k}{\partial t}(x, t)\,\phi(x, t)\,\mathrm{d}x\,\mathrm{d}t = -\int_{3R*R} X_k(x, t)\frac{\partial \phi(x, t)}{\partial t}\,\mathrm{d}x\,\mathrm{d}t$$
(2.51)

$$\int_{3R*R} \frac{\partial X_k}{\partial x}(x, t)\,\phi(x, t)\,\mathrm{d}x\,\mathrm{d}t = -\int_{3R*R} X_k(x, t)\frac{\partial \phi(x, t)}{\partial x}\,\mathrm{d}x\,\mathrm{d}t.$$
(2.52)

It can be shown that:

$$\frac{\partial X_k}{\partial t} + v_i \nabla X_k = 0$$
(2.53)

If f is smooth except at S, then $f \nabla X_k$ is defined via:

$$\int_{3R*R} f \nabla X_k \phi \,\mathrm{d}x\,\mathrm{d}t = \int_{-\infty}^{\infty} \int_S n_k f_k \phi \,\mathrm{d}S\,\mathrm{d}t$$
(2.54)

where n_k is the unit normal vector, exterior to phase k, and f_k denotes the limiting value of f on the phase-k side of S.

It is also clear that X_k is zero, except at the interface. Equation (2.54) describes the behaviour as a function 'picking out' the interface S, and has the direction of the normal, interior to phase k.

Applying a more specific averaging process (time or volume averaging, for example) requires a different set of manipulations regarding the interfacial source terms (Ishii 1975, Nigmatulin 1978).

The volumetric concentration (or void fraction, or relative residence time) of phase k is defined by:

$$\alpha_k = \langle X_k \rangle$$
(2.55)

We note that:

$$\frac{\partial \alpha_k}{\partial t} = \frac{\partial X_k}{\partial t}$$
(2.56)

and

$$\nabla \alpha_k = \langle \nabla X_k \rangle$$
(2.57)

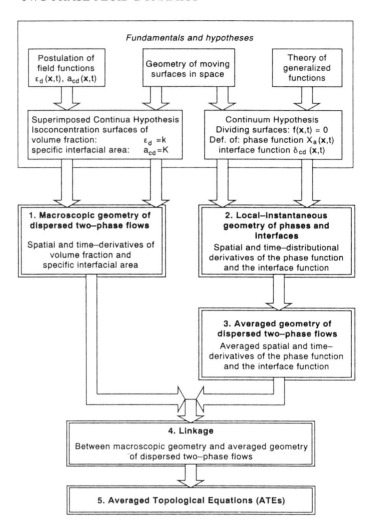

Figure 2.24 Multifluid modelling procedure (Soria and Lara 1992).

The relationship between the volume fraction and the interfacial area is described by so-called *topological equations* (Soria and Lara 1992). According to these authors, the general flow diagram to obtain these equations is presented in figure 2.24.

2.7.2   Three-dimensional model

The general partial differential equations for two-phase non-steady flows in three-dimensional space present mathematical difficulties beyond the present power of theoretical as well as numerical analysis.   However, in many

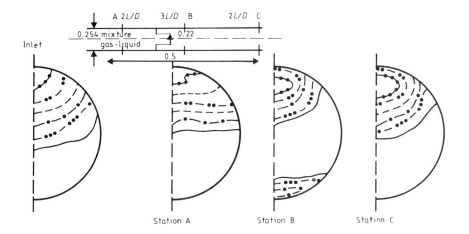

Figure 2.25 Three dimensional void fraction distribution—flow in horizontal channel by Salcudean *et al* (1973); L is entrance length. Iso-voidage lines (%): ●●●● 75, – – – – 70, – · – · – 60, – · · – · · – 50, – · · · – · · · – 20, —— 10.

instances of great interest, simplifications arise, particularly when the number of independent variables reduces to three or two. Such is the case for three-dimensional steady flow, for two-dimensional unsteady flow, for steady flow with cylindrical symmetry, and for non-steady flow with spherical symmetry.

Three-dimensional steady two-phase flow is of practical interest because of its occurrence in heat and mass exchange processes and in Venturi orifices. Experimental results for this kind of flow in a gas/liquid system by Salcudean *et al* (1983) are presented in figure 2.25. A method for the calculation of three-dimensional two-phase flow was presented by Ellul and Issa (1987). This method is based on a multi-dimensional continuum model, in which both phases are enclosed in the same control volume. By solving a set of mass and momentum equations, both void fraction and velocity fields are obtained. The results highlight the potential of the numerical approach which is capable of determining the local details of the flow fields in a wide variety of configurations. The results also reveal the main shortcoming of this method, namely its inability to handle stratification accurately. The remedy for this deficiency is to include models for bubble coalescence and break-up in addition to two-phase turbulence.

2.7.3 Unsteady two-dimensional slug flow

Slug flow is of importance in the oil industry. When a horizontal pipeline is operated in slug flow the prediction of the liquid flow rate at the end of the pipeline is important in the design of separator facilities. During slug flow, the liquid phase exists as aerated liquid slugs and as liquid films, on top of which the gas phase occurs as a bubble. Since most of the liquid exists in the fast

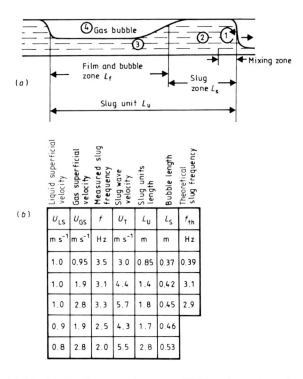

Figure 2.26 (a) Liquid-slug flow—main zones; (b) Experimental results (Kvernvold *et al* 1984), f_{th}—theoretical frequency (Hubbard and Dukler 1966).

moving slugs, an *a priori* knowledge of the average slug length is essential for proper design of separator vessels.

Slug flow can also exist in a gas/solids system. Such flow occurs in dense phase pneumatic conveying pipelines, where the hydrostatic pressure of the air propels the slugs along the pipe.

An idealized slug structure is shown in figure 2.26. This consists of four zones:

- the mixing zone (1)
- the slug body (2)
- the liquid film (or particles settled layer) (3)
- the bubble (4).

The mechanism of the flow is that of a fast moving slug of liquid overriding a slow moving film ahead of it. The slug scoops up the moving liquid film and accelerates it to the wave velocity U_T in the mixing zone. Liquid (or particles) is shed from the tail of the slug to a trailing film which moves more slowly than the slug body. To describe the slug flow, a non-steady three-dimensional

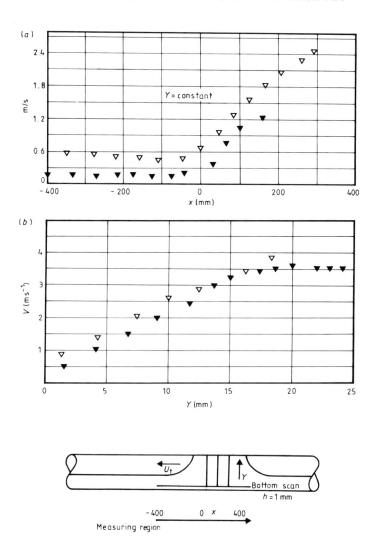

Figure 2.27 (*a*, *b*) Velocity distribution in slug flow. ∇—$U_{LS} = 1.0$ m s^{-1}, $V_{GS} = 1.9$ m s^{-1}, $\blacktriangledown$—$U_{LS} = 1.0$ m s^{-1}, $V_{GS} = 0.95$ m s^{-1}. Results obtained by Kvernvold *et al* (1984). Physical parameters: $P_L = 800$ kg m^{-3}; $P_G = 1.14$ kg m^{-3}; $V_L = 15 \times 10^{-6}$ m^2 s^{-1}; $V_G = 1.57 \times 10^{-6}$ m^2 s^{-1}.

flow model should be taken into consideration. At present, such a model does not exist, and two-dimensional or even one-dimensional models are commonly used.

From the papers by Scott *et al* (1986), Kvernvold *et al* (1984) and Tomita

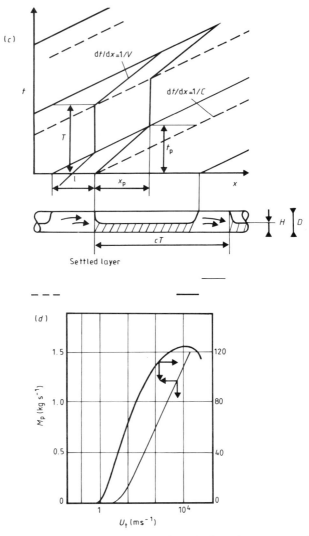

Figure 2.27 (Continued). (*c*) Travelling of particulate slug: —— path of slug rear, ––– path of slug front, — path of particle. (*d*) Mass flow rate of solids and solids loading ratio (M_p/M_e) by Tomita *et al* (1981).

et al (1981), slug characteristics such as slug length, bubble length, slug unit frequency, and velocity field in the slug body are presented in figure 2.27(*a–f*).

2.7.4 Dispersed bubble or particle flow model

The dynamic behaviour of a laminar or turbulent two-phase mixture (gas/liquid or gas/solids) is fundamental to the study of a number of important technical

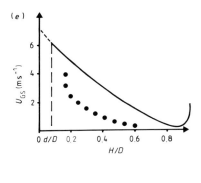

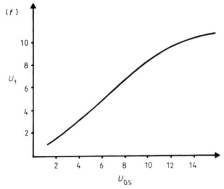

Figure 2.27   (Continued). (*e*) Superficial air velocity versus height of a settled layer in the slug flow regime; the solid line shows the critical air velocity above which saltation occurs over the settled layer of a given height *H*. (*f*) Slug velocity versus air superficial velocity by Tomita *et al* (1981).

problems where an accurate flow and component distribution must be predicted, including such phenomena as heat and mass transfer coefficients, and critical heat flux. One of the most important and yet least understood aspects of two-phase flow is the lateral phase distribution mechanisms which occur. These aspects are often quite pronounced and must be considered in the accurate analysis of heat and momentum transfer for chemical and thermal applications. The lateral phase-distribution describes two-phase flows in one-dimensional macroscale. Nevertheless, to predict this distribution, multi-dimensional effects in microscale should be taken into account as was done by Wang *et al* (1987).

According to Lee and Durst (1982) and Wang *et al* (1987), the lateral velocity as well as void fraction distributions for both bubble and particle flow is given in figures 2.28(*a*–*l*).

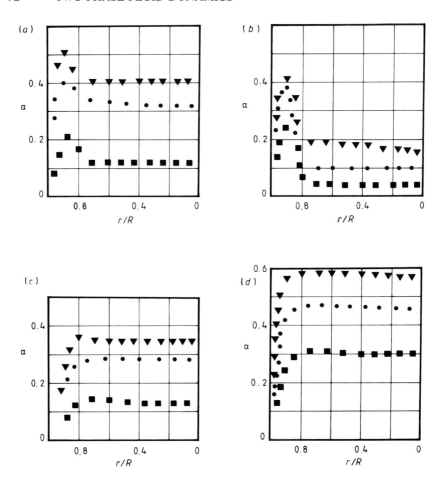

Figure 2.28 Lateral void fraction distribution by Wang *et al* (1987): (*a*) upward flow $U_{LS} = 0.43$ m s^{-1}, (*b*) upward flow $U_{LS} = 0.93$ m s^{-1}, (*c*) downward flow $U_{LS} = 0.94$ m s^{-1}, (*d*) downward flow $U_{LS} = 0.4$ m s^{-1}; ▼—$U_{GS} = 0.10$ m s^{-1}, ●—$U_{GS} = 0.27$ m s^{-1}, ■—$U_{GS} = 0.40$ m s^{-1}.

2.7.5 Annular flow model

This two-phase flow pattern is of importance in the oil industry in gas/condensate and gas/oil systems. It is predicted to exist in risers connected to horizontal flow lines and in vertical risers of gas wells. In this flow regime, the gas stream, with a certain amount of liquid entrained as droplets, flows in the pipe centre, while the remainder of the liquid flows as a thin film in contact with the pipe wall.

The conditions of the fluids in the above applications (gas/condensate or gas/oil systems, with typical pressures of 20 MPa) generally differ widely from

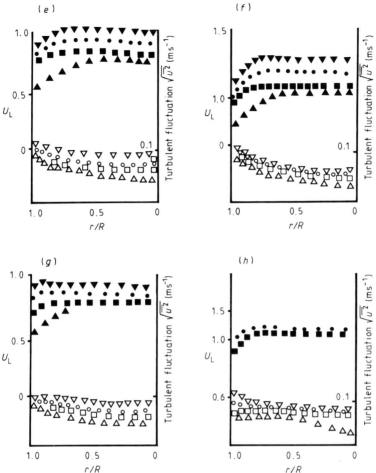

Figure 2.28 (Continued). Lateral velocity distribution by Wang (1987): (*e*) upward flow $U_{LS} = 0.43$ m s^{-1}, (*f*) upward flow $U_{LS} = 0.93$ m s^{-1}, (*g*) downward flow $U_{LS} = 0.71$ m s^{-1}, (*h*) downward flow $U_{LS} = 0.94$ m s^{-1}; ▲, △—$U_{GS} = 0.0$, ■, □—$U_{GS} = 0.10$ m s^{-1}; ●, ○—$U_{GS} = 0.27$ m s^{-1}, ▼, ▽—$U_{GS} = 0.40$ m s^{-1}, where closed symbols represent velocity, and open symbols represent FMS turbulent fluctuation.

those in laboratory experiments at low pressure carried out in vertical flows (section 2.4).

To describe this flow pattern, a one-dimensional model in the direction of the flow can be used. According to Oliemans *et al* (1986) this model is based on the following assumptions:

• the flow is axisymmetrical (e.g.—circumferentially uniform liquid distribution),

• there is upward and developed two-phase flow,

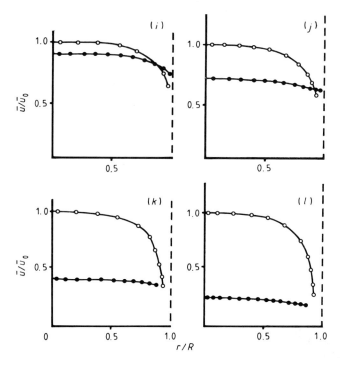

Figure 2.28 (Continued). Lateral velocity distribution: gas–particle flow by Lee and Durst (1982): (*i*) 100 μm particles, $\bar{u}_0 = 5.70$ m s^{-1}, $\alpha = 0.58 \times 10^{-3}$, (*j*) 200 μm particles, $u_0 = 5.84$ m s^{-1}, $\alpha = 0.63 \times 10^{-3}$, (*k*) 400 μm particles, $u_0 = 5.77$ m s^{-1}, $\alpha = 0.72 \times 10^{-3}$, (*l*) 800 μm particles, $u_0 = 5.66$ m s^{-1}, $\alpha = 1.21 \times 10^{-3}$; $\bigcirc$—air velocity, $\bullet$—particle velocity.)

- the liquid droplets in the core travel at the gas velocity (homogeneous flow),
- there is no mass transfer between the gas and liquid phases,
- the inertial forces in the momentum equation are neglected,
- the physical properties (i.e. the liquid and gas densities) are constant.

The correlations for thickness of the liquid layer and the entrainment ratio are given in figure 2.29(*a,b*) (Willets *et al* 1987).

2.8 SUMMARY

The main points in the chapter are

- Flow regime and interfacial effects have a profound effect on the efficiency of conveying systems and of heat/mass transfer processes.

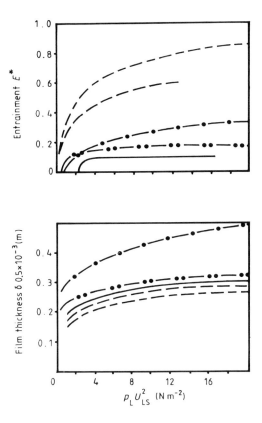

Figure 2.29 Annular flow characteristics: (*a*) entrainment rate. (*b*) film thickness, by Willets *et al* (1987). For the values of $P_G U_{GS}^2$ (N m^{-2}): 222: $-\bullet\bullet-\bullet\bullet-$ (Air/water); 410: ——(Helium/water); 201: $-\bullet-\bullet-\bullet-$ (Air/fluoroheptane); 222: $-----$ (Air/aqueous sulpholane); 231: $-----$ (Air/genklene).

- Flow regimes can be predicted if there is sufficiently accurate knowledge of the operating conditions.

- In many cases operating conditions are not accurately known, so flow imaging is needed to identify the flow regime.

- Information on the sensitivity of the flow regime to variations in process operating parameters can assist in controlling the process in a reliable (e.g. blockage-free) and optimal way.

3

Sensing techniques

3.1 INTRODUCTION

To obtain the data necessary to reconstruct a flow image, similar techniques to those used in medical tomography are needed. The general principle is illustrated in figure 3.1(*a*) where a set of N gamma sources and gamma detectors are used to obtain parallel views of the object space (the pipe). Gamma methods are shown in this illustration, but several other types of sensor could be used (and may be preferable) as discussed in section 3.2.

A number of projections are required in order to enable an image reconstruction computer to determine the contents of the object space. (A projection is defined as one or more volumes wherein the contents are measured by a sensing system.) Thus figure 3.1(*b*) shows a second projection in which the angular orientation between the object space and the gamma beams has been changed. This relative movement is repeated a number of times to give a total of P projections. The accuracy of image reconstruction is dependent on the number of projections P and the number of measurements N in each of these projections (Chapter 4). The number of projections and measurements has a significant effect on the data collection time and on the complexity of the

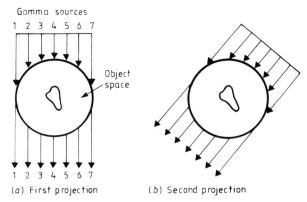

Figure 3.1 Use of projections to form image.

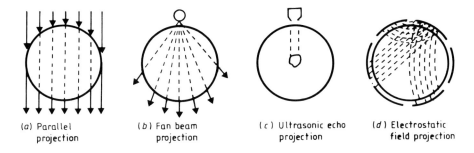

(a) Parallel   (b) Fan beam   (c) Ultrasonic echo   (d) Electrostatic
   projection   projection   projection   field projection

Figure 3.2   Various forms of projection.

transducer system. The effectiveness of the reconstruction algorithm will also depend on P and N (Chapter 4).

Various sensors have been developed for flow imaging and these have various forms of projection associated with them (figure 3.2). Figure 3.2(a) shows the parallel projection, identical to that used in figure 3.1; radiation methods such as x-rays, gamma rays, and light are required to give this type of parallel projection. Figure 3.2(b) shows a fan beam projection method in which one radiation source is associated with a number of detectors to form a single projection. An alternative to the arrangement in figure 3.2(b) is a technique in which a single photon-energy measuring detector and multi-channel analyser is used with a number of radiation sources each having a different energy (Omotosho *et al* 1989). Figure 3.2(c) shows an ultrasonic echo method (section 3.6) where the same sensor can be used for transmission and reception, the ultrasonic echo gives information on the presence and on the distance of the flow component in the object space whereas the other forms of projection only indicate that an object is present along the interrogated beam without reference to its position. Figure 3.2(d) shows the electrostatic field sensing zones which would be encountered when using capacitance sensing plates distributed around the surface of the pipe.

The forms of projection shown in figure 3.2 are only a selection of all those possible when using various configurations of sensors. The wide range of sensing techniques available for flow imaging is shown in table 3.1. In practice each type of sensor requires separate consideration and with certain types of sensor either empirical or computational methods have to be used to determine the actual lines of projection (section 3.4.6). Often the primary requirement is to choose a sensor system which is optimal from the point of view of cost, accuracy, installation complexity, maintenance requirements, etc. At this stage it is then necessary to consider the forms of projection which will be encountered and their effect on the algorithmic reconstruction methods used.

Table 3.1 Sensors for flow imaging.

Principle	Practical realization	Typical applications	General remarks	Remarks re reconstruction algorithms
Modulation of beam of electromagnetic radiation by the dispersed components in the flowing fluid.	Optical techniques	Many 2-component flows where the carrier phase is transparent to the radiation used.	Conceptually simple, high definition possible, fibre optic light guides can simplify optical arrangements. Images of central region poor if second phase concentration is high, due to absorption near walls.	Similar algorithms well established for medical CAT.
	Ionizing radiation—x-rays and γ-rays.	Flows where there is a substantial density difference between the components.	Heavy shielding may be required to collimate beams and for safety. Photon statistical noise limits response time (suitable only for low speed flows unless large sources are used).	

Table 3.1 Sensors for flow imaging (continued).

Principle	Practical realization	Typical applications	General remarks	Remarks re reconstruction algorithms
Reflection of external radiation.	Ultrasonic pulse echo systems.	2-component flows where reflections occur at boundaries e.g. liquid/gas flows.	Ultrasound, about 1 MHz passes through a metal/liquid interface so a 'clip- on' system may be feasible for liquid flow. 'Ringing' of transmitter may cause difficulty in imaging discontinuities close to the pipe walls.	Similar to some NDT and medical applications.
Instantaneous measurement of electrical properties of the flowing fluid.	Electrical capacitance plates on walls of pipe detect the presence of the second component.	Oil/gas, oil/water, gas/solids flows etc.	Inexpensive and rugged. High definition not possible, but good for slug and annular flow, water separation measurements, dune and spiral flow patterns in pneumatic conveyors. Loss of definition near centre of pipe.	Sensor field is affected by distribution of second phase so algorithms must allow for this.
	Conductivity sensing electrodes near wall of pipe.	Water/oil, water/ gas, water/solid flows.	Similar to capacitance, but electrode polarization, greasy deposits etc may need to be considered.	Similar to capacitance. Similar methods used for some medical applications.

3.2 CLASSIFICATION OF SENSORS

There are many ways of classifying sensors but the general division into the two groups of hard field sensors and soft field sensors serves to identify some important characteristics of most of the sensors likely to be encountered (table 3.1).

The reason for this grouping is that it is important to know whether the flowing fluid will have an influence on the field of the sensor. This will have an effect on the choice of reconstruction algorithm (Chapter 4) and on the overall design of the image reconstruction system, with consequent effects on the cost of the system.

3.2.1 Hard field sensors

A hard field sensor is one in which the field of the sensor is not deformed by the flow. Typical hard field sensors use gamma rays, x-rays or light beams. These all fall into the category of sensors in which external radiation is modulated by the flowing fluid (Beck and Plaskowski 1987). Ultrasound (section 3.6) is sometimes regarded as a 'hard field' but this is not always the case (as discussed below).

The general principle of a hard field sensor is illustrated in figure 3.3 where a beam of incident intensity I_0 is being transmitted through the fluid in the plane perpendicular to the axis of the pipe. This energy will pass through a volume of fluid in which it is scattered and absorbed to give an emergent intensity I, given by:

$$I = I_0 \exp(-\mu D) \tag{3.1}$$

where D is the separation between the source and the receiver and μ is the total attenuation coefficient per unit length of fluid:

$$\mu = \mu_s + \mu_a \tag{3.2}$$

where μ_s is the scattering attenuation coefficient and μ_a is the absorption attenuation coefficient.

The attenuation coefficients μ_s and μ_a depend upon many factors (type of fluid and discontinuities in the fluid, energy of radiation source, concentration and distribution of discontinuities, wavelength of radiation, etc). In particular, equation (3.1) is true only when the wavelength of the radiation is much less than the size of the flow component elements in the pipe (always true for gamma rays and sometimes for light). However, when ultrasonic radiation is used, the simple relation (3.1) is not true because of scattering by the flow component elements and the vectorial (phase) modulation by turbulence in the conveying fluid (Beck and Plaskowski 1987).

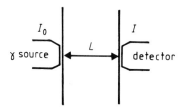

Figure 3.3   Radiation absorption.

Table 3.2   Electric field parameters

	Electrostatic	Magnetostatic	Current Flow
Field intensity	$E = -\nabla\phi$	$H = -\nabla\phi_m$	$E = -\nabla V$
Potential	ϕ electric potential	ϕ_m magnetic potential	V voltage
Flux density	$D = \epsilon E$ electric flux density	$B = \mu H$ magnetic flux density	$J = \sigma E$ current density (Ohm's law)
Component parameter	ϵ permittivity	μ permeability	σ conductivity

3.2.2   Soft field sensors

In a soft field sensor the initial sensor field is distorted by the electrical properties of material in the image space. In this category one can include all impedance-based sensors (e.g. capacitance, resistance). For example the electrostatic sensing field in capacitance sensors changes according to the distribution and permittivity of the components in the flow.

Soft field sensors are taken here to include any technique by which measurements of voltages and current conditions on the boundary of a region can be used to determine the spatial distribution of conductivity or permittivity in the region. Theoretically the sensor field can be described by one of the sets of parameters given in table 3.2 (Seager *et al* 1987).

Images produced using data from soft field sensors are of low spatial resolution. This is because of the sensor field distortion discussed above and consequently increasing the number of electrodes will not give a proportional improvement in image quality. The current (or field) can be driven between each of N electrodes in only $(N - 1)/2$ independent ways, hence in total there are

$$M = N(N - 1)/2 \qquad (3.3)$$

independent measurements, which also imposes a limitation on the image resolution (figure 4.9).

The images obtained by the various soft field techniques have common characteristics. The quality of reconstruction is poorer further away from the electrodes. Thus the accuracy and spatial resolution of the image is worse near the centre of a pipe when wall-mounted sensing electrodes are used. Noise in the measurements also causes the image quality to deteriorate, and this effect is worse at points far away from the electrodes. The situation is different when using hard field sensors, e.g. gamma ray techniques, where a change in gamma ray absorptivity caused by an object on the path of the radiation does not depend so critically on the position of that object in the path.

The soft field methods of measurement are particularly attractive for flow imaging because they do not involve the use of gamma radiation (safety problems), light (obscuration problems), or ultrasound (low sonic speed and sensor complexity problems). However, the soft field methods are essentially nonlinear, i.e. the effect of an object in the field depends upon its position relative to the electrodes, and therefore it is important to obtain high accuracy data from the transducers to enable the nonlinear effects to be compensated for in the reconstruction algorithm (Chapter 4).

Increasing the number of electrodes can improve both the resolution and the accuracy, but there are practical problems associated with this, including those concerning low signal level with small electrodes, cost and complexity when more electrodes are used, and the effect of 'field spread' at points distant from the electrodes which means that the resolution cannot be increased by simply increasing the number of electrodes (section 3.4.6).

3.3   MEASUREMENT FOR FLOW IMAGING

The specific requirements for flow imaging will now be considered. Suppose we have measurements denoted by vector y, then the aim of image reconstruction is to solve:

$$y = \mathbf{A}x \tag{3.4}$$

where y is a vector (dimension m) which contains data measured externally by sensors, x is a vector (dimension n) representing the set of unknown coefficients (e.g. pixel grey levels) characterising the flow component over a pipe cross-section, $\mathbf{A}$ is a $m \times n$ matrix containing known sensitivity factors.

More details are given in Chapter 4, but here we must consider that we have a y vector which contains data from the sensors. It indicates the total effect of the flowing material in the sensing field. Matrix $\mathbf{A}$ encodes the contribution that any pixel makes towards the observed (measured) total value. Therefore x represents the local parameters of the flow (density, attenuation, permittivity, etc) in the sensing field.

Several measurement techniques can be used for flow imaging (table 3.1), but here we will consider only electrical field methods (sections 3.4 and 3.5) and ultrasonic methods (section 3.6). These examples, which are based on instruments developed in the UK, lead to general design principles which the reader can use to solve a particular problem.

3.4 CAPACITANCE SENSORS—A MAJOR CASE STUDY

3.4.1 Introduction

In recent years, there has been a growing application of capacitance techniques to the measurement of component concentration in multi-component flows. Some laboratory studies have been carried out to investigate the relationship between the measured sensor capacitance and component concentration in various flows. These include the measurement of gas fraction in gas/oil mixtures (Abouelwafa *et al* 1980a), water content in oil (Hammer 1983) and solid particle concentration in powder/gas streams (Irons and Chang 1983). The capacitance measuring equipment used in the above applications was often general purpose commercial capacitance meters, laboratory impedance analysers, etc. There are also a few reports on transducers specifically designed for pneumatically conveyed solids flow measurement (Huang *et al* 1988a) and gas void fraction measurement in gas/liquid 2-phase flows (Hammer 1983, Auracher and Daubert 1985, Lucas 1987).

Some specific problems associated with the use of capacitance measuring techniques for flow imaging, such as the effects of stray capacitance, electrostatic interference generated by the fluid, and measurement sensitivity distribution, have been investigated by Huang (1986).

3.4.2 The fundamental requirements of a capacitance-based concentration measurement system

The principal problem in concentration measurement is achieving a good signal-to-noise ratio. By signal we mean the sensor capacitance change caused by the concentration change of a relevant component of the flow. Some applications require sensor capacitance changes of about 0.001 pF or less to be detected, such a small signal can be easily masked by various noise sources. There are several factors which can cause unwanted changes in the capacitance transducer output and hence can be regarded as noise sources.

Firstly, the baseline drift of capacitance measurement circuits due to temperature and other effects may be significant. This is particularly relevant to the design of on-line industrial transducers which are often required to operate for long periods without recalibration.

Secondly, the distribution pattern of the component to be measured in the flow is not always constant. A typical example is in the flow of pneumatically

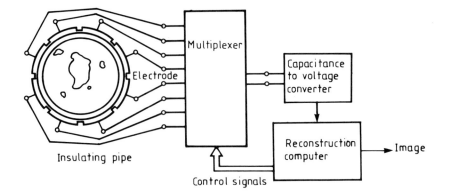

Figure 3.4 Capacitance imaging system.

conveyed solids where, as the solid loading increases, the solid distribution in the pipe gradually varies from the homogeneously dispersed state to the roping state, in which the solids travel as a helical rope along the pipe line, resulting in very irregular distributions over the pipe cross-section (Green 1981). If the measurement sensitivity distribution of the capacitance sensor is not homogeneous over the pipe cross-section, then the measured capacitance is not only dependent on the solid concentration but also on its distribution in the pipe, and this will cause unpredictable measurement errors.

Thirdly, some fluids, especially pneumatically conveyed solids, carry strong electrostatic charges which behave as noise sources and will disturb the output of a capacitance measuring circuit unless effective noise rejection measures are taken.

Therefore, capacitance sensors for multi-component flow processes must be designed with minimal baseline drift, a homogeneous measurement sensitivity distribution, a high sensitivity to component concentration changes, and high immunity to electrostatic interference. Some ways of doing this are described by Huang *et al* (1988a).

3.4.3   A capacitance flow imaging system for non-conducting fluids

In pipelines where the continuous component is an insulating fluid, capacitance sensing techniques using stray-immune transducers (Huang 1986) provide a simple and economic means of implementing flow imaging systems, for applications such as the visualization of component distribution in multi-component flow pipelines (Huang *et al* 1988b, 1989).

An 8-electrode capacitance flow imaging system is shown in figure 3.4. The sensitive regions of each of the capacitance electrode pairs are confined to relatively small areas of the pipe cross-section, enabling an acceptable image resolution to be obtained. The four basic patterns of the sensitivity distribution are shown in figure 3.5. These patterns can be made to rotate around the pipe

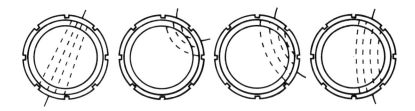

Figure 3.5 Capacitance imaging system—typical sensitivity distribution patterns.

centre by the multiplexer (figure 3.4). According to the number of possible combinations of electrodes, 28 different measurements can be obtained in 1 scan for an 8-electrode system. The data collection system consists of a multiplexing circuit and a capacitance-to-voltage transducer. The multiplexing circuit selects 2 of the 8 electrodes, and the capacitance transducer measures the capacitance between them. In a complete operation cycle, the capacitance of electrode pairs 1–2, 1–3, 1–4, 1–5, 1–6, 1–7, 1–8, 2–3, 2–4, 2–5, 2–6, 2–7, 2–8, 3–4, 3–5, 3–6, 3–7, 3–8, 4–5, 4–6, 4–7, 4–8, 5–6, 5–7, 5–8, 6–7, 6–8 and 7–8 are measured sequentially, and this produces a total number of 28 independent measurements. The number of possible combinations of 2-electrode pairs in an N-electrode system M is given by equation (3.3). The 28 measured data, whose values depend on the concentration and distribution of one component in another, are fed into the computer which reconstructs the cross-sectional image of the component distribution from the measured data. The computer also controls the data collection processes such as multiplexing and A/D conversion. The capacitance-to-voltage transducer is a stray-immune type, i.e. it is sensitive only to the capacitance between the selected electrode pair, and insensitive to stray capacitance between the measuring and redundant electrodes. In addition, the redundant electrodes are always held at the virtual earth potential, forming guard rings to the measuring electrodes. These techniques enable the measurement sensitivity to be focused in a relatively narrow area between the selected electrode pair.

In order to define the reconstruction algorithm it is necessary to know the sensitivity distribution of the sensing field. This can be calculated by using the finite element method (Xie *et al* 1989a). Figure 3.6 shows the distribution patterns for measurements made on electrode pairs 1–2 and 1–3. The sensitivity distributions of other electrode pairs can be obtained by rotating the 4 standard patterns around the centre of the pipe.

Figure 3.6 shows that the sensitivity distributions are not homogenous over the pipe cross-section. There is an area between the measuring electrode pair where the sensitivity is positive, whereas in other areas in the pipe the measurement sensitivity is small and in some areas the measured capacitance responds negatively to a dielectric increment (Huang 1986). Therefore, for a given pair of electrodes, there could be many permittivity distributions, for

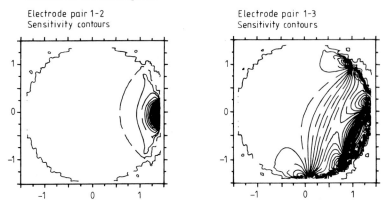

Figure 3.6   The measurement sensitivity distribution between electrode pairs 1–2 and 1–3.

example a smaller object at a more sensitive point and a larger object at a less sensitive point, which would result in the same measured capacitance value. To simplify the reconstruction problem, it is assumed that any change in a measured capacitance results from a homogeneous change in the permittivity over the entire positive sensing area. In reconstructing the image, the positive sensing areas are given grey levels which depend on the measured capacitance change. Section 3.4.6 describes how finite-element CAD methods can provide detailed information of electrode field design.

3.4.4   Practical limitations

The capacitance method gives relatively low resolution images but these are adequate for many two-component flow measurement applications (Chapter 6). This is mainly because the dimension of the sensor electrodes is quite large, and this results in some large-sized pixels (figure 4.9). Any object with a size smaller than the size of the pixel cannot be correctly reconstructed. Increasing the number of electrodes will improve the resolution, but the measurement sensitivity will decrease, making the detection of small dielectric changes in a pipe more difficult.

   The standing capacitance of the electrode pair in such a system is independent of the pipe diameter, but dependent on the number of the electrodes and the electrode length in the flow direction, which should be kept short if it is desired to reduce the spatial averaging of the component distribution in the axial dimension (this reduction may be needed when measuring the velocity profile by pixel cross-correlation, Chapter 1). With the electrode length equal to 110 mm (about 1.2 times the pipe diameter), the 8-electrode system shown in figure 3.4 has a minimum standing capacitance of about 0.3 pF between electrodes separated by a pipe diameter, e.g. between electrodes 1–5, and this requires a transducer with a long term accuracy of better than ±0.0015 pF to detect a 1% change in sensor capacitance. Further increase in the number of electrodes will necessitate higher sensitivity and stability of the capacitance transducer.

Unlike medical imaging, where the objects to be imaged are stationary, 2-component flows often travel at high speeds, say several metres per second. This requires flow imaging systems to have high data collection and reconstruction speeds. The time needed to collect one measurement value depends mainly on the settling time of the capacitance-to-voltage transducer, which in this system is about 500 μs. The total of 28 measurements thus takes 14 ms to complete, which is sufficiently short for a large number of applications.

In calculating the measurement sensitivity distribution of the capacitance sensors (figure 3.6), it has been assumed that permittivity is homogeneous over the entire cross-section of the pipe. However, the component distribution of 2-component flow is inhomogeneous. This causes distortion of the field equipotentials between the capacitance electrodes, and results in a distorted image. The extent of the distortion depends on the difference between the permittivities of the 2 components and the inhomogeneity of the distribution.

Various methods have been proposed to reduce the image distortion (Bair and Oakley 1992, Chen *et al* 1992). For example, an iterative approach can be used. This involves recalculating the capacitance field using the distorted image generated in the first iteration, reconstructing the image using the newly calculated sensitive area, and then recalculating the capacitance field using the new image, and so on, until the undistorted image is approached. Obviously the iterative approach will increase the computer loading.

3.4.5   Electronic design of capacitance sensors for flow imaging

In section 3.4.3 it was stated that the measuring circuit should be relatively immune to stray capacitances between the electrodes and ground. It is also convenient if the circuit is switched digitally so that the scanning arrangement shown in figure 3.4 can be readily implemented. A sensor based on the charge transfer principle (figure 3.7) satisfies both criteria. By using fast CMOS switches for transferring charge the operating frequency, which is programmable, can be as high as 5 MHz so the quadrature capacitive reactance can be measured with minimum error caused by the in-phase dielectric loss (i.e. so that loss of sensitivity due to resistive leakage is minimized). This feature is needed when imaging 'lossy' dielectrics, such as an oil flow with sea water contamination (Huang 1986).

The basic circuit of the sensor devised by Huang (1986) is shown in figure 3.7(*a*), and the timing sequence of the CMOS switches S_1 to S_4 in figure 3.7(*b*). Basically, the measurement consists of charging the unknown capacitance C_x to a known voltage $+V_c$ and then discharging it. This charge/discharge cycle is repeated under the control of a digital clock signal with a programmable frequency f. The successive discharging current pulses

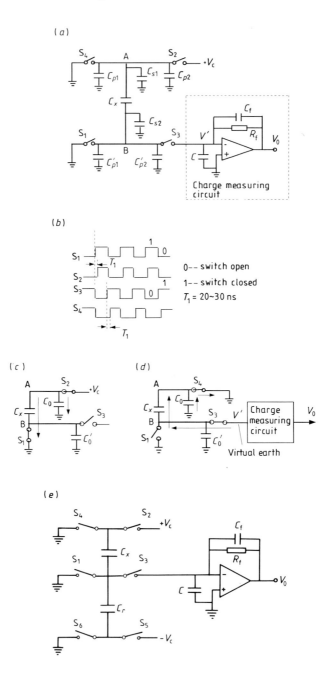

Figure 3.7 The stray-immune charge transfer circuit.   (a) circuit diagram, (b) switching control signals, (c) charging process, (d) discharging process, (e) differential configuration to reduce drift.

of C_x are measured by a charge detector based on an operational amplifier, figure 3.7(a).

Because the slew rate of the amplifier is not fast enough for the input to hold at virtual earth when the discharge pulse arrives, the discharging current pulses flowing through it generate transient voltage spikes at the input of the amplifier. To reduce consequential errors a smoothing capacitor C ($C \gg C_x$) between the amplifier input and ground, is used to absorb the transients and ensure a stable virtual earth potential at the input of the charge detector without affecting the measured charge.

By choosing a large integration time constant $T_f = R_f C_f$ and a large C value, the charge measuring circuit produces a dc output voltage proportional to the unknown capacitance and good measurement accuracy can be maintained at high switching frequencies without special requirements for the bandwidth of the operational amplifier.

In a typical charge/discharge cycle, when the clock signal goes high, switch S_1 is closed to connect electrode B of the sensor C_x (known as the detecting electrode) to earth. After a very short time delay T_1, S_2 is closed to charge the other electrode A (the source electrode) to $+V_c$ (figure 3.7(c)). C_0, the overall stray capacitance effectively connected to electrode A, including those of the switches, C_{p1} and C_{p2} and that between electrode A and the screen, $C_s 1$ (figure 3.7(a)), is also charged via S_2 to V_c, whereas the overall stray capacitance connected to electrode B, C_0, is discharged to zero potential by S_1. When the clock goes low, S_1 and S_2 open before switch S_3 closes connecting electrode B to the detector input which is held at virtual earth potential. After a very short time delay T_1, S_4 closes to discharge both C_x and C_0 to earth (figure 3.7(d)). Only the discharge current of C_x will flow through the detector and that of C_0 will not. Therefore, the effects of the stray capacitance C_0 are eliminated from the measurement.

Since C_0 was at earth potential before S_3 closes, the charge drawn by C_0 from the detector after S_3 closes is (figure 3.7(d))

$$Q' = C_0'V'. \qquad (3.5)$$

As there is no current leakage path (shunting resistance) between the detecting electrode B and earth, the total amount of charge flowing through the detector during the discharging interval is

$$Q = C_x V_c + C_0'V'. \qquad (3.6)$$

Compared with the charging voltage V_c (15 V in this design), V' is very small, typically a few millivolts for a JFET input operational amplifier. Therefore, the detector is much more sensitive to change in C_x than to that in C_0' since

$$\frac{\delta Q/\delta C_0'}{\delta Q/\delta C_x} = \frac{V'}{V_c} = 0.0002 \qquad (3.7)$$

and the influence of the stray capacitance C_0' is significantly reduced.

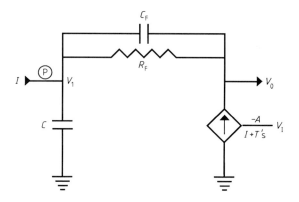

Figure 3.8   Transposed version of charge measuring circuit in figure 3.7(a).

3.4.5.1   Theoretical modelling and analysis.
A. *The transfer function of the transducer.* For convenience in the following analysis figure 3.8 shows a transposed version of the charge measuring circuit shown in figure 3.7.

The transfer function allowing for the amplifier phase inversion is

$$H(s) = \frac{-V_0(s)}{I(s)} \tag{3.8}$$

where s is the Laplacian operator. The op-amp gain characteristics give

$$V_0 = \frac{-A}{1 + T's} V_I \tag{3.9}$$

where A is the open loop steady state gain and T' the open loop time constant, hence

$$V_I = -\frac{1 + T's}{A} V_O. \tag{3.10}$$

Equating the currents out of node P to zero

$$-I + sCV_I + \left[\frac{1}{R_f} + sC_f \right] [V_I - V_0] = 0 \tag{3.11}$$

Re-arranging (3.11)

$$I = V_I \left[\frac{1 + R_f(C + C_f)s}{R_f} \right] - V_0 \left[\frac{1 + R_f C_f s}{R_f} \right] \tag{3.12}$$

Substituting for V_I from (3.10)

$$I = -\left[\frac{1 + T's}{A} \left(\frac{1 + R_f(C + C_f)s}{R_f} \right) + \frac{1 + R_f C_f s}{R_f} \right] V_0 \tag{3.13}$$

$$= -\left[1 + \frac{1}{A} + \left(R_f C_f + \frac{R_f(C + C_f)}{A} + \frac{T'}{A} \right) s \right.$$
$$\left. + \frac{R_f(C + C_f)T'}{A} s^2 \right] \frac{V_0}{R_f}. \tag{3.14}$$

Assume $A \gg 1$ and $A R_f C_f \gg R_f C_f + R_f C + T'$, then

$$H(s) = -\frac{V_0(s)}{I(s)} \approx \frac{R_f}{1 + R_f C_f s + \frac{R_f(C+C_f)T's^2}{A}} \qquad (3.15)$$

The charge in the capacitance C_x is effectively a series of impulses of magnitude

$$Q = V_c C_x \qquad (3.16)$$

at a repetition rate of f. Hence the current has a DC component

$$I = f V_c C_x \qquad (3.17)$$

plus sinusoidal components at the angular frequency $2\pi f$ rad s^{-1} and all the higher harmonics.

The detector is designed to have a 'low' pass bandwidth which filters out these sinusoids whilst allowing the signal frequencies to pass, thus the response of the transducer to a capacitance signal $C_x(s)$ is given by

$$V_0(s) = H(s)I(s) \qquad (3.18)$$

$$= \frac{R_f f V_c C_x(s)}{1 + R_f C_f s + \frac{R_f(C+C_f)T'}{A}s^2} \qquad (3.19)$$

In the steady state this reduces to

$$V_0 = R_f f V_c C_x. \qquad (3.20)$$

The transducer sensitivity is given by

$$\frac{\Delta V_0}{\Delta C_x} = R_f f V_c. \qquad (3.21)$$

B. *Baseline and sensitivity drift.* To reduce the sensor baseline drift caused by variations in charging voltage V_c, switching frequency f, component values, etc (see (3.20)), a differential configuration has been devised (figure 3.7(*e*)) with a reference capacitor C_r charged to a negative voltage $-V_c$ while C_x is charged to $+V_c$. Switches S$_5$ and S$_6$ control the charge and discharge of C_r and their operations are synchronized with those of S2 and S4.

Equation (3.20) for the differential configuration is rewritten as

$$V_0 = f V_C R_f (C_x - C_r) + e + i R_f \qquad (3.22)$$

where e is the input offset voltage of the amplifier and $i R_f$ the voltage error due to the input offset current. The change in measured capacitance due to parametric variations is obtained by differentiating (3.22) and dividing the result by the sensor sensitivity (3.21) giving

$$\Delta C_m = \Delta C_x - \Delta C_r + (C_x - C_r)\left(\frac{\Delta f}{f} + \frac{\Delta V_c}{V_c} + \frac{\Delta R_f}{R_f}\right) + \frac{\Delta e + \Delta i R_f}{R_f f V_c}. \quad (3.23)$$

Equation (3.23) shows that the output drifts caused by $\Delta f/f$, $\Delta V_c/V_c$, and $\Delta R_f/R_f$ are eliminated if

$$C_x = C_r \quad\quad\quad (3.24)$$

The effects of Δe and Δi are reduced by increasing the sensor sensitivity, as shown in (3.23).

Baseline drift is minimized by controlling the drift sources. ΔC_r can be minimized if a ceramic capacitor with a low temperature coefficient is chosen (in some cases a parallel combination of ceramic and mica capacitors with positive and negative temperature coefficients can be used to reduce drift). Low drift operational amplifiers are selected for the measuring circuit. For instance, a typical JFET-input op-amp has an input offset voltage which varies with temperature but a negligible input offset current change with temperature. When the sensor sensitivity $f V_c R_f$ is chosen to be $1\ \mathrm{V\,pF^{-1}}$, the maximum change in output corresponding to Δe is equivalent to less than 0.0001 pF change at the input capacitance sensor for a $\pm 10°\mathrm{C}$ change in ambient temperature. By using a quartz crystal oscillator (100 ppm), a precision voltage regulator (10 ppm/°C) and 0.1% precision resistors (15 ppm/°C), the relative output drift, $\Delta C_m/(C_x - C_r)$, caused by $\Delta f/f$, $\Delta V_c/V_c$, and $\Delta R_f/R_f$, is easily confined to within 0.1% for a $\pm 10°\mathrm{C}$ change in ambient temperature.

C. Selection of the parameters f, V_c, R_f and C_f. The above analysis shows that the values of the switching frequency f and the charging voltage V_c should be as high and as stable as possible. In practice, the highest usable value of V_c depends on the semiconductor switches (15 V for CMOS analogue switches). The maximum f value is limited by the settling time of the transient voltage oscillations generated at the detecting electrode due to the high speed switching operation. Increasing the feedback resistor value R_f, increases the measurement sensitivity of the sensor and its sensitivity to noise interference, e.g. the electrostatic charge generated in solids/air 2-component flows (Huang 1986), and is chosen to be between 10 kΩ and 100 kΩ. The value of C_f is selected to provide the required signal bandwidth (equation (3.19)). A wide bandwidth is needed for fast flows, but can only be achieved at the cost of a worse signal-to-noise ratio.

3.4.6   Computer aided design of capacitance electrode systems

The electrode field pattern has a profound effect on the quality of the image, with uniform and well collimated fields giving better images than those possible

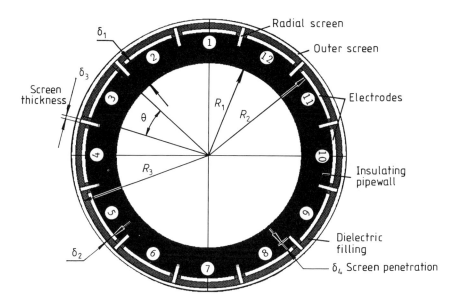

Figure 3.9 Cross section of a 12-electrode capacitive sensor (not in scale) (from Khan and Abdullah 1992).

with non-uniform and dispersed fields. Now we will consider a CAD method for the optimal design of flow imaging electrode structures, which exploits the potential of the 'stray free' sensor (section 3.4.5) to give symmetrical field sensitivity between electrodes and to tolerate the close proximity of earth screens without losing sensitivity. The work described in this section was carried out at City University (Khan and Abdullah 1991, 1992) and is seen as fundamental to establishing cost-effective design procedures for the wide variety of types and sizes of electrode systems needed for flow imaging.

As a case study we will consider the CAD and performance analysis by the finite element method (FEM) of a 12-electrode capacitance system for 2-component flows. The choice of 12 electrodes gives a reasonable compromise between sensitivity and image resolution—section 3.4.6.2.

Figure 3.9 shows the cross section of a 12-electrode system consisting of 12 capacitance electrodes mounted symmetrically on the outer surface of the insulating wall of a section of the pipeline. The radial screens between electrodes reduce the high capacitance between neighbouring electrodes and force the electric field towards the central region. The earthed outer screen with inner radius $R3$ acts as a shield to make the whole system immune to external fields. The inner radius, $R1$, of the pipe wall is fixed but the outer radius, $R2$ and hence the wall thickness, $\delta 1 = R2 - R1$, can vary. The space between the pipe wall outer surface and the outer screen is filled with dielectric material to insulate the electrodes from the screen. The quality of the reconstructed image

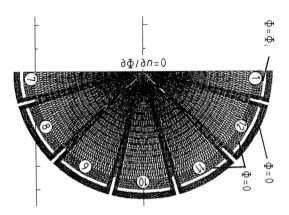

$\partial \Phi / \partial n = 0$

Figure 3.10   Finite element model of a 12-electrode capacitive sensor.

depends on the uniformity of the sensitivity distribution which is determined by the following system parameters: the number of electrodes N, electrode angle Θ, pipe wall thickness $\delta 1$, the distance between the electrodes and outer screen $\delta 2 = R3 - R2$, radial screen thickness $\delta 3$ and its penetration depth inside the pipe wall $\delta 4$. These parameters must be optimized to get the best system performance.

3.4.6.1   Finite element modelling of a 12-electrode capacitance system.   By assuming that fringing field effects due to finite electrode lengths are negligible and that the flow component distributions do not change spatially along the axial direction of the electrodes as with core flow and annular flow, then the electrode system in figure 3.9 can be simulated using the 2D finite element (FE) mesh pattern, shown in figure 3.10. For radially symmetrical flows the electric field distribution between electrodes is symmetrical which enables only half of the model to be used. This increases the accuracy of FE solutions since a finer mesh can be generated for the same number of nodes.

For flow component distributions with permittivity given by $\epsilon = \epsilon(x, y)$ the electric field in the electrode system can be calculated (assuming free charge distribution) by solving the equation in terms of electrostatic potential, $\Phi = \Phi(x, y)$

$$\nabla \cdot (\epsilon \nabla \Phi) = 0. \tag{3.25}$$

This equation can be solved by finite element (FE) methods using, for example the Vector Fields software PE2D, under the boundary conditions shown in figure 3.10 which gives potential values at nodes from which the field component vectors E and D can be calculated. As the accuracy of any FE solution depends on various aspects of discretisation special care has been taken to minimize discretization errors by using a large number of elements, typically nearly 10 000. A consistent mesh pattern has been used for all models to ensure the comparability of the results. After solving (3.25), capacitances between electrode pairs are found by calculating the total charge on the detecting electrode (by integration using Gauss's Law).

To analyse the effects of geometric parameters the following system performance parameters are used: (a) standing capacitance per unit length of electrode, C_{0ij}, i.e., the capacitance between electrode pair $i-j$ when the pipe is empty (relative permittivity $\epsilon = \epsilon_0$); (b) absolute sensitivity of the system, $\Delta C_{ij}^k = (C_{ij}^k - C_{0ij})\ell$ where, C_{ij}^k = capacitance between electrode pair $i-j$ when only the kth elementary finite element region inside the pipe has permittivity $\epsilon = \epsilon t$; and ℓ is the length of the electrodes; (c) relative sensitivity of the system, $S_{ij}^k = (C_{ij}^k - C_{0ij})/(C_{0ij})^*(A_0/A_k)$ where A_0 is the cross-sectional area of the pipeline and A_k is the area of the kth finite element region.

3.4.6.2 Quantitative design.

One of the main parameters of multi-electrode capacitance systems is the number of electrodes, N. As shown in Chapter 4 the image resolution of the system increases as the number of independent capacitance measurements, M increases. Since the number of independent measurements (section 3.2.2) is $M = N(N-1)/2$ the higher the N the better the image resolution. However, as N increases the data acquisition and image reconstruction times also increase. This is undesirable as a flow imaging system must have a fast response due to the high speed of flow components. Moreover, as N increases the electrode angle Θ decreases, which increases the minimum detectable void fraction α (cross sectional area of an elongated bubble flow relative to A_0) of the system. For example, for a gap of 5 between neighbouring electrodes, the values of Θ for a 12- and an equivalent 24-electrode system are 25° and 10°, respectively. As shown in figure 3.11 (Khan and Abdullah 1991) such a 12-electrode system can detect void fractions one third the size of those detectable by a 24-electrode one. Capacitances C_{0ij} and C_{ij}^k decrease as N increases, so a larger N requires more sensitive measuring circuits. Considering all these aspects $N = 12$ seems to be a reasonable compromise. However, for slow flows N could be increased to obtain higher resolution images (see discussion on signal-to-noise ratio in section 3.4.5).

For a given electrode system the maximum and minimum capacitances between adjacent electrodes (i.e. the capacities with the pipe full and with the pipe empty) C_{max} and C_{min} depend on Θ. In order to increase measurement accuracy it is desirable to minimize the ratio $K_c = C_{max}/C_{min}$, provided that the ratio between non-adjacent electrodes is kept high. Figure 3.12 shows how K_c changes with Θ for 8 and 12-electrode systems. This factor must be taken into consideration when selecting the electrode gap because of its effect on Θ.

K_c is also influenced by parameter $\delta2$, the distance between electrodes and the outer screen. As shown in figure 3.13, K_c increases as $\delta2$ increases, which suggests that $\delta2$ should be as small as possible. However, a small value of $\delta2$

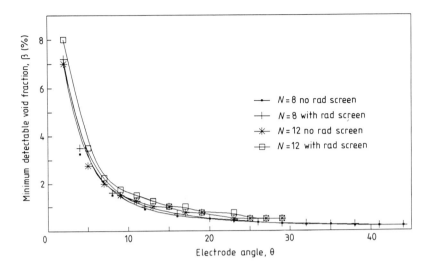

Figure 3.11   Variation of minimum detectable void fraction with electrode angle (from Khan and Abdullah 1991).

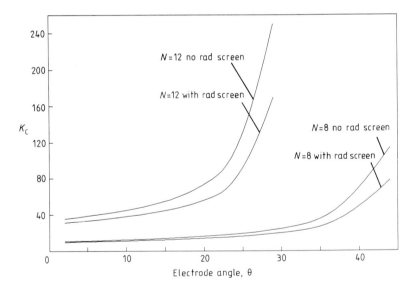

Figure 3.12   Variation of K_c with electrode angle.

increases the electrode capacitance to earth, which, if excessively high, could cause errors in the measurement (Huang 1986).

Radial screens are useful features in multi-electrode systems. For obvious geometric reasons, the capacitances between an active electrode and its closest

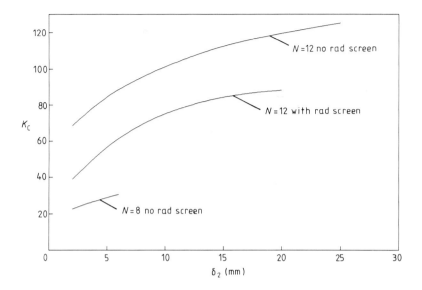

Figure 3.13   Variation of K_c with δ_2.

neighbours is much higher than those between an active and the farthest electrodes. The main purpose of radial screens is to decrease the electric field between the active electrode and the neighbouring electrodes thus decreasing K_c and so increasing measurement accuracy.   The shielding effects of the radial screen between electrodes 1 (active) and 12 can be seen clearly by comparing equipotential plots for the 12-electrode system in figures 3.14(a,b). In figure 3.14(a) the screen does not penetrate the pipe wall ($\delta 4 = 0$) but in figure 3.14(b) it does ($\delta 4 = 5$ mm). Quantitatively, these effects can be seen from the curves of K_c versus Θ and K_c versus $\delta 2$ shown in figures 3.12 and 3.13. The introduction of radial screens is thus justified. However, as shown below for a given $\delta 1$ their effectiveness depends mainly on $\delta 4$ and $\delta 3$ which, in addition to shielding, affect the sensitivity distribution inside the pipe.

The standing capacitances between electrodes change with radial screen thickness $\delta 3$ although other results show that it does not significantly affect the overall sensitivity distributions. Figure 3.15 shows how capacitance $C_{01,12}$ between electrode pair 1–12 varies with $\delta 3$ for a range of $\delta 4$. It is evident from figure 3.15 that from a shielding point of view a shorter and thicker screen can be as effective as the longer and thinner one. From mechanical considerations a shorter $\delta 4$ is preferable.

Figures 3.16 and 3.17 show some of the effects of $\delta 1$ and $\delta 4$ on sensitivity distribution between electrode pairs 1–7 and 1–12 for a 12-electrode system. The sensitivities $S_{1,7}$ and $S_{1,12}$ have been calculated from the respective changes in capacitance due to the change in permittivity of elementary regions along the

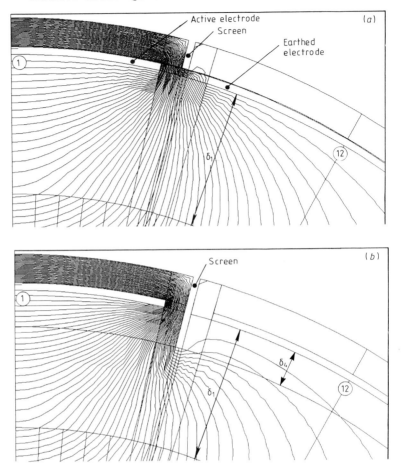

Figure 3.14 Equipotential plots showing effect of radial screen penetration (*a*) zero penetration, (*b*) 5 mm penetration.

radius (radial regions) in figure 3.16 and on an arc (regions on arc) close to the pipe wall (figure 3.17). In figures 3.16 and 3.17 the numbers (1, 2, 3, ...) along x-axis correspond to positions of these regions which are specified in figure 3.18. In figures 3.16(*a*) and 3.16(*b*) the 'C's along the x-axis represent the position of the elementary regions at the centre of the pipe. Careful analysis of figures 3.16 and 3.17 shows that pipe wall thickness $\delta 1$ plays an important role in sensitivity distributions and that a more uniform distribution can be achieved by using a thicker pipe wall. This has useful implications for electrode design as discussed in the next section.

3.4.6.3 Electrode fabrication and application. When applied to oil/gas flows the design method described in section 3.4.6.2 leads to the use of electrodes

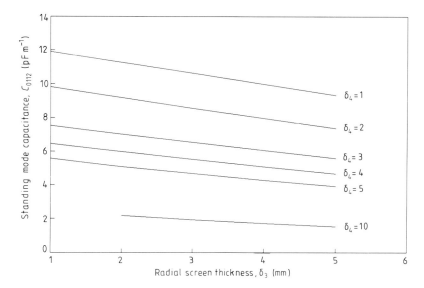

Figure 3.15 Variation of capacitance between adjacent electrodes of 12-electrode sensor with screen thickness $\delta 3$ for various screen penetrations $\delta 4$ in mm.

where the gap between the electrode face and the fluid, is about 15 mm for a 150 mm diameter pipeline. This gap should be filled with a material having a dielectric constant approximately similar to that of the oil, which can be achieved by using any of several types of plastic material (Perspex, PTFE etc). This is used in the design of the electrode system shown in figure 3.19, where a simple electrode unit can be slid onto a plastic pipe with a suitable wall thickness (section 3.4.6.2). Details of the slide-on electrode system are shown in figure 3.20 and a photograph of electrodes for flame imaging is shown in figure 3.21. The versatile slide-on system can be readily moved from one pipeline to another for experimental work at low pressure. It can be made suitable for high-pressure installations by fitting it into a robust enclosure.

3.4.7 Capacitance sensor control and interfacing to an image reconstruction computer—case study

This case study is based on a 12-electrode system (section 3.4.6) and uses parallel data collection to minimize the collection time and to maximize the signal/noise ratio (figure 3.22) A detailed description of the electronic configuration is given by Huang *et al* (1992a) and of the software control and reconstruction by Xie *et al* (1992).

An array processor (Transputer) system is used for image reconstruction. This provide a useful facility for increasing speed by adding more Transputers without software changes, and it can be interfaced to a PC for program development.

3.4.7.1 Interfacing. As shown in figure 3.22, a CO11 Transputer link adaptor

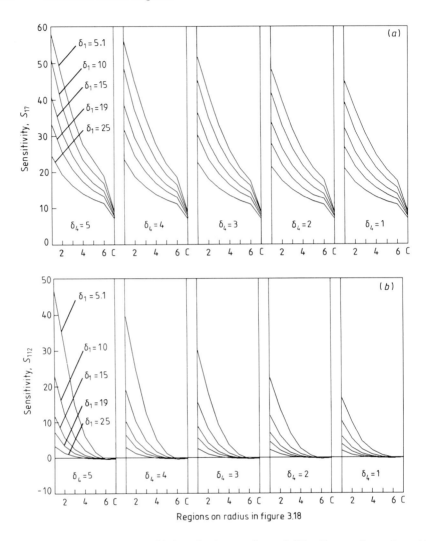

Figure 3.16 Variations in sensitivity of (*a*) opposite and (*b*) adjacent electrodes with regional positions for various screen penetrations ($\delta 4$ in mm) and wall thicknesses ($\delta 1$ in mm).

(from INMOS) can be used to provide an interface between the sensor electronics and the Transputer system for image reconstruction. The link adapter is built into the sensor electronics, and is connected with the Transputer system via two serial communication links, one input and one output. Control codes from the Transputer are converted by the CO11 into parallel outputs, and the parallel output from the A to D converter is converted into a serial output and sent to the Transputer system. The functions of the control codes include:

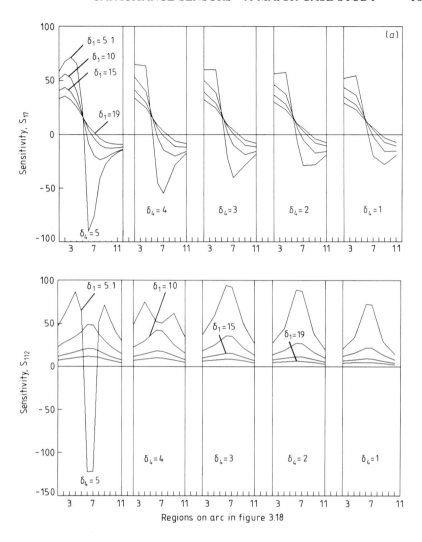

Figure 3.17 Variations in sensitivity of (*a*) opposite and (*b*) adjacent electrodes with regional positions for various screen penetrations and wall thicknesses (δ_1 in mm).

1. selection of operating mode, measurement or baseline checking,
2. selection of electrode status (source, detecting or idle),
3. selection of measurement channel outputs for A to D conversion,
4. zero balance,
5. gain control,
6. A to D conversion,
7. switching frequency selection (625 kHz for low-sensitivity or 1.25 MHz for high-sensitivity measurement).

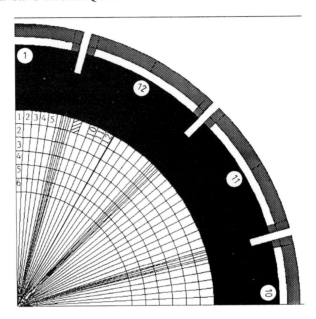

Figure 3.18 Showing positions of elementary regions on radius and on arc, used for evaluating sensitivity in figures 3.16 and 3.17.

3.4.7.2 Sensor control. A problem associated with the capacitance imaging system is that the standing capacitance values and sensitivity of different electrode pairs can be very different. For instance, for a 12-electrode system, the standing capacitance value (when the pipe is empty) between electrodes 1 and 2 is about 100 times greater than that between electrodes 1 and 7, and even for similar pairs, such as 1–2 and 1–12, the measured standing values may be different due to the manufacturing errors of the electrodes. This causes difficulties for the capacitance measuring circuit which has a relatively limited dynamic range. To avoid saturation when measuring the capacitance between adjacent electrodes, the gain of the circuit has to be kept at a value that is too low for measuring the diagonally separated electrodes.

The solution to this problem is to use a programmable reference voltage to balance the different standing values of the measurements (zeros) and a programmable gain amplifier to satisfy the different gain requirements (figure 3.22). For a 12-electrode system there are 66 independent measurements (equation (3.3)). A calibration procedure has been designed to determine the 66 zeros and the required measurement gains (figure 3.23) and it should be performed before the measurement and imaging process is started. During the calibration, the sensor pipe is first filled with the component of lower permittivity (e.g. air for a gas/liquid 2-component process) and the 66 capacitance values are measured and stored as the zeros. Then the pipe is filled with the component of higher permittivity and the gains of the circuit are selected to enable each

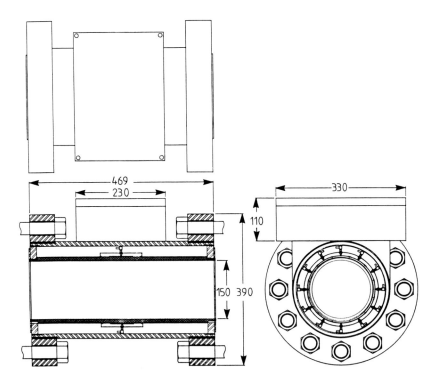

Figure 3.19 Imaging electrodes slide over plastic pipe. Pressure-safe tube outside (Tealgate Ltd).

different measurement to achieve the required output level. The 66 values of gain thus obtained are then stored in the Transputer system. During the measurement process (figure 3.24) each time a measurement channel is selected the stored zero and gain values for this particular configuration are sent from the Transputer to the offset and gain control units to preset the appropriate zero balance and gain before A to D conversion starts. The offset and gain control units can receive 10 bit digital codes and this allows a fine adjustment of the corresponding parameters.

Thus the data required for image reconstruction (section 4.4) are generated. The actual capacitance measurements used for image reconstruction are the differences between the capacitances of each electrode pair with material present and their standing capacitances (i.e. the capacitance of the electrode pair when the sensing volume is filled with the component of lowest permittivity).

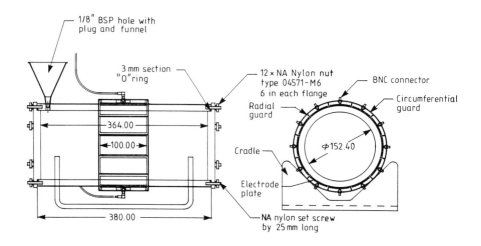

Figure 3.20   6″ (152.4 mm) 'Slip-On' sensor mounted on calibration stand.

Figure 3.21   Electrical capacitance tomography system being used for flame imaging (measuring chemi-ionization of flame. Electrodes supplied by Shell Thornton Research Centre, tomography system supplied by Process Tomography Ltd.

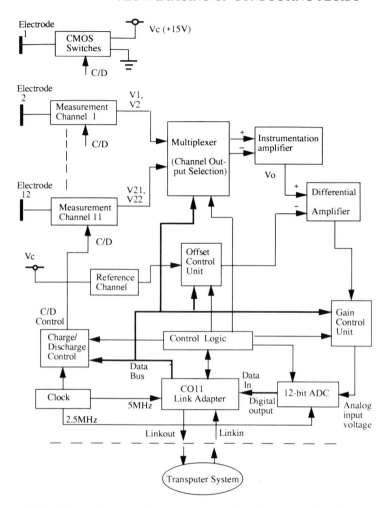

Figure 3.22   Block diagram of the sensor control and transputer interface system.

3.5   ELECTRICAL IMPEDANCE TOMOGRAPHY FOR FLOW IMAGING OF CONDUCTING FLUIDS—AN OUTLINE

In section 3.4 we described a capacitive imaging system suitable for use with electrically non-conducting fluids, which are frequently encountered in the oil industry, and with gas/solids pneumatic conveying. In many other processes the fluids are water-based and therefore electrically conducting, for these processes electrical impedance tomography (EIT), in which the resistive component of impedance is measured will operate analogously to the electrical capacitance tomography method for non-conducting systems.

EIT was developed for medical applications in the early 1980s (Barber

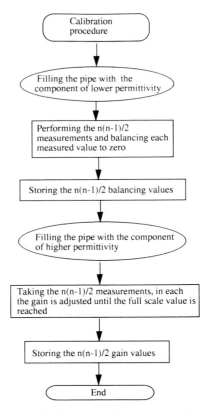

Figure 3.23   The calibration procedure.

and Brown 1984). More recently this method has been extended to process applications and here we will describe the process tomography system developed by a group led by Dickin at UMIST (Wang *et al* 1992).

Referring to figure 3.25, the EIT system consists of non-intrusive electrodes on the inner wall of the vessel, a data acquisition system (DAS) and an image reconstruction system (IRS). A common practice is to inject the current signal into one pair of electrodes and measure the voltage developed on the remainder of the electrodes. This process is then repeated for other pairs of current injection electrodes. By separating current injection and voltage measurement, errors which would be caused by the voltage drop at current injection electrodes (due to polarization effects) are avoided.

The electrodes can be small (Wang *et al* 1992) so 16 or even 32 electrodes can be accommodated around a flow pipe, hence in principle a rather higher resolution should be attainable with electrical impedance tomography compared with that of capacitance tomography which requires much larger electrodes.

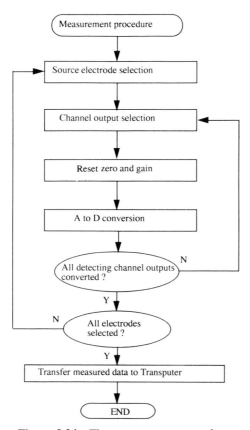

Figure 3.24   The measurement procedure.

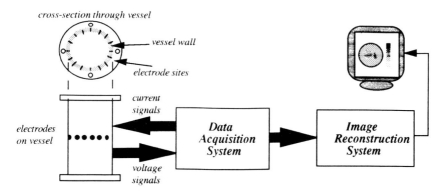

Figure 3.25   Schematic block diagram of the EIT instrument.

3.5.1 Electrodes and interfacing to the data acquisition system

Point electrodes on the periphery of the pipe section in figure 3.25 are positioned equidistantly at fixed locations in such a way that they make electrical contact with the fluid. They are connected to the data acquisition system by cables of short length. Practical studies indicate that the use of tri-axial cable provides better signal-to-noise ratios than co-axial cable. The outer sheath of the triaxial cable is earthed in the same way as the co-axial cable to act as a guard, whilst the middle sheath is coupled to a voltage follower to reduce the effective cable capacitance. The inner core is capacitively coupled to the input of the voltage buffer (section 3.5.2.2).

The materials used for the electrodes depends largely on the process application. The material must be more conductive than the fluids being imaged, otherwise problems will occur due to contact impedance. Typically, the electrode material is brass, stainless steel or silver palladium alloy which is commercially available in bolt- or screw-form and can often be threaded into the vessel wall. The number of electrodes (N) can be varied. However, it must be noted that the accuracy of measurements is a function of N, whereas the time taken to acquire the data and reconstruct the image is approximately proportional to N. From equation (4.19) for $N = 16$ electrodes, the number of unique measurements $M = 104$. However, for twice the number of electrodes ($N = 32$) M increases to 464.

3.5.2 The data acquisition system

The data acquisition system (DAS) is responsible for obtaining the quantitative data describing the state of the resistivity distribution inside the pipeline. The data has to be collected quickly and accurately in order to track small changes of resistivity in real-time thus enabling the reconstruction algorithm to provide an accurate indication of the true resistivity distribution. We will now consider the critical subsystems of the DAS shown in figure 3.26.

3.5.2.1 Constant current generator. The accuracy of the whole EIT measurement system is principally determined by the accuracy of the current driver stage which is composed of a master oscillator and a voltage-to-current converter. The oscillator serves as a frequency and amplitude reference for all of the current source channels and as a switching function for the demodulator stage. A suitable oscillator is a Datel ROJ-1K hybrid integrated circuit sine-wave signal generator having an amplitude accuracy of 0.5% at 10 kHz. The frequency can be adjusted from 1 kHz up to 100 kHz by a computer-controlled resistor network enabling interrogation of the pipeline contents over a range of different current injection frequencies. Such flexibility will optimize visualisation of objects exhibiting different frequency response characteristics and thus allow the DAS to be 'customized' for specific applications.

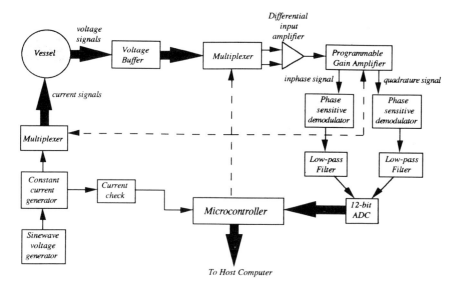

Figure 3.26   Components of a typical EIT data acquisition system.

The sine-wave voltage output from the oscillator is fed into a voltage-to-current converter (often referred to as a voltage controlled current source—VCCS). Current is used in preference to voltage as the electrical 'probe' due to the variation of contact impedance between the electrode and the fluid inside the vessel.

The VCCS circuit employs 2 operational amplifiers and an analogue switch arrangement in the feedback path of one of the op-amps in order to minimize the VCCS's output admittance Y_0. This arrangement enables the VCCS to maintain a constant amplitude over a wide range of resistance loads (i.e. better than 0.5% peak-to-peak amplitude variation over a load range of 100Ω to 2.5 kΩ).

3.5.2.2 Differential input amplifier.   The output from the voltage measuring multiplexers is routed into a differential input amplifier, which amplifies the potential difference between the two input voltage signals. The important attribute of such an amplifier is its ability to 'reject' common-mode signals such as extraneous electrical noise appearing on both input lines. The common-mode rejection ratio (CMRR—units: dB) is a figure of merit for differential amplifiers and a figure of 110 dB at 50 kHz can be obtained by repeatedly changing over the inputs to the amplifier, at a high frequency compared with the response time, to negate the effects of input impedance imbalance between the two inputs. The sine-wave output of the differential amplifier is then fed into a programmable gain amplifier (PGA) to accommodate the wide dynamic range of voltage signals obtained from the electrodes.

3.5.2.3 Demodulation and filtering. A phase-sensitive demodulator (PSD) is employed after the PGA stage to demodulate the voltage signal. Analogue linear multipliers perform demodulation on the real and imaginary parts of the voltage signal using two reference voltages from the master oscillator, namely a sine-wave (in-phase) and a cosine-wave (quadrature) waveform. After low-pass filtering the PSD outputs to remove harmonics of the fundamental frequency the resulting signals are given by

$$V_{in-phase} = 0.5V_x A \cos\theta \qquad (3.26)$$

$$V_{quadrature} = 0.5V_x A \sin\theta \qquad (3.27)$$

where A is the amplitude of the master oscillator output, and Θ is the phase difference between the unmodulated signal (V_x) and the reference signal. The availability of both the real and imaginary components of the voltage signal provides additional measurements for applications where impedance spectroscopy (frequency dependent impedance of materials) may in future be used to identify specific components in a multicomponent mixture.

3.5.2.4 Digital control. From the preceding subsections it is clear that many of the circuit stages require one or more digital control lines to function. The stages utilizing such lines are summarized below:

(i) Input/output multiplexers—for N electrodes, $4\log_2 N$ control lines are required.

(ii) Programmable gain amplifier—number of lines required depends upon the number of amplification stages, usually $\times 10$ and $\times 100$, thus 2 lines are needed.

(iii) Programmable frequency—depends on number of desired frequencies, for example 3 frequency settings can be handled by a 4:1 multiplexer with 2 control lines.

(iv) Programmable current amplitude depends on number of desired amplitudes. 3 amplitude settings can be handled by a 4:1 multiplexer with 2 control lines.

(v) Programmable low-pass filter—similar to (iii).

(vi) Change-over switches (to improve CMRR)—depends on number of differential input amplifier stages. If a single differential input amplifier is multiplexed only 2 control lines are necessary.

(vii) ADC output and control (12 output lines and 2 control/status lines).

(viii) Current check switch.

Not all of the above stages will need use of the control lines at the same time. By using octal latches it is possible to time-multiplex the lines in order to maximize their utilization. Thus, it is quite feasible to control all of the above programmable 'features' for a 16-electrode system using only 24 digital input/output control lines.

3.6 ULTRASONIC SENSORS—AN OUTLINE

3.6.1 Introduction

The electrical field methods of tomographic imaging described in sections 3.4 and 3.5 use sensors which are principally sensitive to permittivity (capacitance tomography) and to resistance (EIT). An alternative to these is the ultrasonic transducer which is sensitive to density changes and has the potential for imaging component flows such as oil/gas/water mixtures which occur frequently in the oil industry. In such cases, ultrasonic techniques could be used to image the gas component (large density differences), while capacitance techniques could be used to image the water component (large permittivity difference); thus providing individual images of the gas and water components flowing in an oil well, riser or pipeline.

This section is based on the work done by Gai (1990) at UMIST. Ultrasonic techniques are used to make measurements in many areas of science and technology (Lynnworth 1989). The majority of publications on ultrasonic tomography are concerned with medical or NDT (non-destructive testing) applications. Asher (1983) summarizes the use of ultrasonic sensors in the chemical and process industries, while a good introduction to NDT is the book by Silk (1984). In Plaskowski *et al* (1987) the authors point out the special difficulties in multi-phase flow imaging. The basic theory in ultrasonics has been well developed and ultrasonic instrumentation is widely used (Crecraft 1983). In 1984 Schueler reviewed the history of acoustic imaging and the fundamentals of digital ultrasonic imaging (Schueler *et al* 1984). At the same time an interesting paper was published by Schafer and Lewin (1984) in which the authors thoroughly described the design, construction and development of the front-end hardware of a digital ultrasonic imaging system. The recent trend in front-end hardware has been new sensor material development and further implementation of integrated circuit techniques.

Apart from searching for new materials, many researchers have been investigating improved methods of acoustical impedance matching between the media in order to reduce the attenuation (Jayet *et al* 1983). Another active area is to shape or guide the ultrasonic beam by using acoustic lenses (Dines and Gross 1987). Phased array sensors (Von Ramm and Smith 1983) are attracting more attention because the associated electronics has improved greatly in the past 10 years, although the narrow beam angles from a phased array are not suitable for imaging high speed flows, where a wide beam is needed to cover the pipe area (section 3.6.5).

3.6.2 Limitations of ultrasound in flow imaging

Although the use of ultrasound to produce images is not new, there are a number of problems that need to be considered if the technique is to be successfully applied in flow imaging applications (Gai 1990).

First of all, ultrasound is highly attenuated in most materials, and the differences of attenuation (characterized by impedance) in different states of material (solid, liquid, gas) are so large that it is extremely difficult to model its behaviour at state interfaces. For multi-phase flow, many state interfaces can sometimes exist. This limits the image resolution.

Secondly, ultrasound propagation is frequency dependent. In other words, at different frequencies its behaviour changes in a given medium. This can be attributed to the sizes of the particles or bubbles being comparable with the wavelength of the ultrasound, therefore undesirable scattering, reflection and mode changing can occur under certain conditions. In multi-phase flow the size and shape of components make it extremely difficult to design transducers so the spurious effects can be reduced by processing the output data, especially when unpredictable particle sizes exist in the flow or when the flow has rapidly changing patterns (Chapter 2).

Thirdly, ultrasound travels too slowly to give high resolution images of fast moving flows in large pipes. For example, consider an oil/gas flow with the inner diameter of the pipe as 150 mm. The speed of ultrasound in oil is about 1.5 $km\,s^{-1}$ or 1.5 mm μs^{-1}. The journey time needed for an ultrasonic reflection to be received from an object at the far side of the pipe is 200 μs. Let us assume there are 8 individual sensors arranged in the cross-section plan of the pipe, so one complete scan takes 1.6 ms without switching time and without data processing. An object will be reliably distinguished only if the time is present in the cross-sectional imaging plane is greater than the time for two complete sensor scans. Thus for an object 15 mm in length (10% of the pipe diameter) the highest allowable flow velocity for reliable identification is 4.7 $m\,s^{-1}$. This is a basic limitation in ultrasonic flow imaging systems.

3.6.3   Interaction between ultrasound and 2-component flow

Ultrasound does not appear to be suitable for the detection of one liquid in another (e.g. oil in water or water in oil), since it passes from one liquid to the other with little reflection. There is, however, a large difference of acoustic impedance between gases (including steam) and liquids. This causes strong reflection of ultrasonic signals from the interface between these components.

Voids in flowing liquids can assume a large variety of shapes (Chapter 2). The interface will in general be a curved surface. To simplify our analysis we will idealize it as a spherical surface.

An ultrasonic wavefront which encounters a bubble is reflected from the curved surface as a diverging wavefront. Only a small part of the energy is reflected back towards the sensor which originated the wavefront. A similar or smaller amount will also be reflected to other sensors which may be within view. In any case, the reflected signal will come from only one part of the bubble surface. It will not give a reliable indication of the size of the void.

A small amount of acoustic energy enters a bubble and some of this reflects from its back surface and re-enters the liquid at the front of the bubble. However, the interface losses are so high that this signal is usually undetectable.

If the bubble is similar in size to the wavelength of the ultrasound in the liquid, little or no reflection will occur, but the wavefront will be perturbed (refracted).

The result of these interactions is that if an ultrasonic pulse radiated from a sensor encounters a sizable bubble, one echo pulse will be returned to the sensor, reflected from the nearest surface of the bubble. The time elapsed between the transmitted pulse and the echo return is

$$t = 2d/c \qquad (3.28)$$

where d is the distance from the sensor to the object and c is the speed of sound. The elapsed time for reflected pulses to reach other sensors is:

$$t_2 = (d_1 + d_2)/c \qquad (3.29)$$

where d_1 and d_2 are the distances from the reflecting surface of the bubble to the sending and receiving sensors, respectively.

3.6.4   Limitation of conventional ultrasonic sensors

Commonly available ultrasonic sensors radiate a beam which may be essentially parallel or may be focused to a point. To detect voids anywhere within a pipe it would be necessary to use a large number (n) of such sensors distributed evenly around the pipe wall. To enable the signal sources to be identified, an individual sensor would have to be pulsed at a time interval greater than

$$\tau = 2nD/c \qquad (3.30a)$$

at a frequency less than

$$f = c/2nD \qquad (3.30b)$$

where D is the pipe diameter.

For a 100 mm diameter pipe, a satisfactory coverage would be achieved with 16 sensors having 15 mm beam widths. With $c = 1500$ m s^{-1} one complete scan would take 2.1 ms.

To ensure that a bubble is interrogated by each sensor at least once during the cycle, it must not move more than 1/2 of the beam diameter, or 7.5 mm during the cycle. This sets the maximum flow velocity at 3.6 m s^{-1}.

To ensure that voids occurring at any location within the pipe cross-section are viewed by as many sensors as possible, it is essential that the sensors have radially diverging beams, the widest practical beam angle appears desirable.

To ensure that voids are viewed for an equal time by each sensor, the beam widths in the axial direction should be as uniform as possible. The combination of this requirement and radial divergence results in the need for a 'fan-shaped' beam pattern (section 3.6.5).

The time resolution required from the signal conditioning and image reconstruction system will depend upon the number of data display elements representing the pipe cross-sectional area.

With p^2 square pixels representing a pipe of diameter D the length 's' of the side of a pixel $= D/p$. This means we can represent an object's position to a discrimination of $\pm s/2 = \pm D/2p$. It would seem reasonable to measure the object's position to the same discrimination [or say to twice this accuracy $\pm s/4$].

Assuming that the same sensor is used for transmitting and receiving then the echo delay time $T_e = 2d/c$, $d =$ distance sensor to object. The distance measurement discrimination is $\Delta d = \pm c \Delta T_{e/2}$, equating this to the representation discrimination:

$$\Delta d = \frac{c \Delta T_e}{2} = \pm \frac{s}{2} = \pm \frac{D}{2p} \qquad \text{i.e. } \Delta T_e = \frac{D}{pc}$$

or, for the more stringent criterion $\Delta d = \pm s/4$,

$$\Delta T_e = \frac{D}{2pc}. \tag{3.31}$$

3.6.5 Ultrasonic sensors for flow imaging

To construct the best possible image from the limited number of interrogations in each measurement round, it is essential to obtain as much information as possible from each interrogation. As discussed in section 3.6.4 a sensor having a fan-shaped beam pattern will cover a wide angle of the flow media in the image plane. This is a method which gains more information in an individual interrogation at the expense of lower sensor sensitivities. Some researchers have used this arrangement in ultrasound tomography (Gai *et al* 1989a, Gai 1990) and have constructed multi-segmented versions which are capable of detecting the directionality of the reflectors and have investigated the application of such sensors to reflection mode ultrasound flow imaging using the system shown in figure 3.27.

It is natural to form the image plane by evenly spacing the fan- shape beam sensors around the pipe unless the imaging system is dedicated to a specific kind of flow, for example, gas/liquid stratified flow. In such special cases the sensors can be arranged in a diametrically symmetrical pattern. Generally speaking, if more sensors are provided, more measurements can be obtained and a better quality of image can be expected.

To construct the primary sensor, two conventional methods can be considered. One is to insert fan-shape beam sensors (Gai *et al* 1989a, b) invasively around the pipe. The other is to clamp planar sensors onto the pipe from outside with

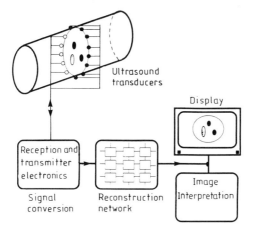

Figure 3.27   Ultrasound flow imaging system.

an acoustic coupling between the sensor surface and the outside pipe wall. The areas of the pipe wall where the sensors are mounted can be flattened to fit the planar surfaces of the sensors. Inside the pipe, facing the outside sensors, convex part-cylindrical acoustic lenses (Tarnolzy 1965, Lemons and Quate 1979) can be cemented onto the pipe inner wall to re-shape the field pattern of the planar sensors into fan-shape. The material of the lenses should have lower acoustic impedance or at least the same as that of the pipe, preferably around the geometrical average of that of the pipe and of the liquid component of the flows (Lynnworth 1965).

Because invasive sensors actually contact the flow inside the pipe for obvious reasons they are not favoured by most industries. The 'clamp-on' technique avoids penetration of the pipe wall and has been successfully used in some industrial flow meters (Lynnworth 1981, 1982). However, if the clamp-on technique is applied to flow imaging, where many sensors are required at different angles around the pipe, the mechanical complexity becomes overwhelming. If the complication of internal lenses is added to this, it is apparent that any advantage of a clamp-on structure is lost. The only practical alternative is the use of an integral transducer/pipe structure

3.6.6   An integral ultrasonic sensor/pipe structure for flow imaging—case study (Gai 1989)

The design of a non-invasive primary sensor for flow imaging applications will now be discussed.

This particular design combines a conventional 'clamp-on' technique common in many ultrasonic flowmeters, with acoustic lenses and utilizing the properties of piezoelectric polymer/copolymer films to produce a novel type of imaging sensor (Gai *et al* 1989b).

Mechanical structure. The aim of this design is to simplify the structures suggested by conventional techniques and thereby to remove the substantial drawbacks associated with them. The philosophy adopted is to integrate all the components of the sensing unit into one independent, modular and robust structure.

The integral structure comprises a length of pipe, the inner and outer walls of which have been formed into special shapes. Specifically, the inner wall has projections which are effectively acoustic lenses and the outer wall has either flattened or concave areas at the positions corresponding to the acoustic lenses. These areas are used for mounting the piezoelectric sensors as explained in the next subsection.

Acoustic lenses can be fabricated as part of the pipe, evenly spaced around a cross-section of pipe inner wall, by machining or casting or some other method. Thus one of the two discontinuity layers is eliminated and so is the difficult task of attaching acoustic lenses. The prefabricated structure has no removable parts and can withstand any kind of flow conditions. Both the shapes and positions of the acoustic lenses must be well defined to produce the expected field pattern and image plane.

Because in this application the dimension of the acoustic lens is many times the wave length, in order to produce a wide angled fan-shape beam pattern, the shape of the lens has to be semi- cylindrical or convex. The field pattern will largely resemble the shape of the lens surface. In the basic structure of this design the shape of the front face of the acoustic lens is convex semi-cylindrical, with both ends streamlined gradually to the pipe wall so as to reduce the obstructing effect on the flow. Figure 3.28 shows the basic structure.

Each individual lens changes the ultrasound beam of the sensor element located outside the pipe into a fan-shape beam pattern which covers a wide angle of the flow media inside the pipe section. Thus when one sensor element transmits a pulse, all the other sensors elements within the coverage angle will receive the transmitted signal. To ensure that the maximum amount of information is received from each transmitted pulse it should be received by as many other sensors as possible. Therefore each sensor should be capable of transmitting and receiving over a wide angle.

Choice of active element and sensor/pipe integration. Ultrasonic sensors usually employ piezoelectric ceramics, polymers or copolymers (Lovinger 1985).

Piezoelectric ceramics usually have high sensitivity. Complex sensor surfaces can be fabricated and polarized in the desired direction or active mode. This group of materials can withstand high temperature, for example, well over 250°C, without losing piezoelectric properties. Their disadvantages are (1) they have poor damping characteristics, that is, they ring for a long time after being pulsed; and (2) they are brittle and cannot readily be made to conform to a curved surface like the pipe wall.

Piezoelectric polymer films in general have lower sensitivity and lower

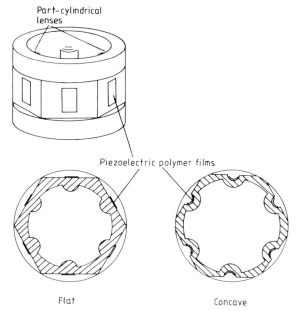

Part-cylindrical lenses

Piezoelectric polymer films

Flat   Concave

Figure 3.28   Basic sensor structure.

critical temperatures, compared with the piezoelectric ceramics. However, the excellent damping characteristics and flexibility make them very attractive in many applications. The piezoelectric properties have been much improved especially with recent developments in copolymers, (Ohigashi 1985, Ohigashi *et al* 1984) and these films have become more important as materials for ultrasound sensors (Galletti *et al* 1988)

Piezoelectric polymer and copolymer films seem to be the most suitable materials, because they are flexible they can be easily attached to the pipe wall. Because they have very good damping characteristics the sensors can be air-backed, that is, the attached active element can be simply exposed to the atmosphere. Figure 3.29 shows the arrangement of the active elements.

The piezoelectric film is cut and cemented to the prepared areas of the pipe. If the pipe section is metal, it can serve as a common terminal of the sensors. The layer of cement securing the active element should be as even and, in particular, as thin as possible (Silk 1984), provided adequate strength is maintained. In the case of a metal pipe, conductive adhesive can be used to eliminate one lead to each sensor.

The sensor leads can be attached by conductive adhesive, for example, silver loaded epoxy. In the case of a non-metallic pipe, a small tag can be provided with the leads attached on both sides, as shown in figure 3.30. A clamp near the element is needed to restrain the leads. The pipe section, with internal lenses and sensor elements attached, serves as the integral front-end sensing unit of a flow imaging system.

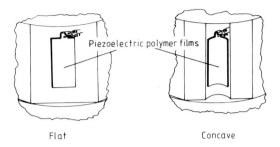

Figure 3.29   Arrangement of active elements.

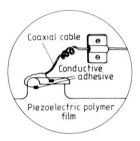

Figure 3.30   Arrangement of sensor leads.

The effect of wave propagation on sensor design. When a sensor transmits a pulse, its acoustic lens will spread it into the liquid contained in the pipe section. If the liquid does not include particles which scatter the ultrasound, the pulse will be transmitted over a wide angle to reach all the acoustic lenses and the intervening segments of the pipe wall which are within the coverage angle. Because of the impedance mismatch between the pipe material and the liquid, when the pulse encounters the pipe wall it divides into a transmitted fraction, which enters the pipe wall, and a reflected fraction, which returns through the liquid. The reflected pulses will travel around the pipe interior, dividing repeatedly at the solid–liquid boundaries, until they are attenuated to a negligible level.

    If there is a discontinuity in the liquid within the coverage angle of the transmitter, an air bubble for example, some of the acoustic energy will be reflected or scattered and some sensors, including the transmitting one, may receive a reflected signal caused by the air bubble. On the other hand, some sensors may be shielded by the bubble and consequently fail to receive the directly transmitted pulse. This picture of transmission, reflection and particularly multiple reflection, seems very complicated at first glance, but by proper timing and monitoring the signal amplitudes, most of the information

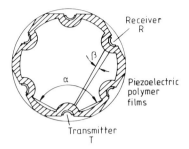

Figure 3.31 Wave propagation in pipe.

useful for multicomponent flow measurement and image reconstruction can be obtained as we will see in Chapter 4. This sensor-pipe arrangement can be used for both transmission mode and reflection mode ultrasonic flow imaging measurements.

Because of the irregular shape of the pipe section, the mode conversion of the compressional waves generated by an individual sensor is very complicated. Different modes of waves will occur and propagate in all directions in the pipe. However, the diverse acoustic energy transmitted through the pipe wall is at very low levels and the majority of acoustic energy goes through the acoustic lens in both transmission and reception situations. This can be shown in figure 3.31 where the major part of the acoustic energy generated by the transmitter T will go into the ultrasound transparent media through angle α which is the coverage angle of the individual sensor. The major part of the acoustic energy picked up by the receiver R will come through angle β, where the surface of the receiving lens is substantially perpendicular to the incident waves. Thus there is not much acoustic energy from the transmitter transmitted around the pipe wall to the receivers. Furthermore, the compressional piezoelectric element is very insensitive to waves in other modes. For example, for compressional wave and surface waves of the same energy level, the pick-up of the latter by a compressional sensor arranged perpendicular to the structure surface is usually less than 5% of the former.

The frequency of the ultrasound used should generally be decided by the resolution requirements. Low frequency should be used whenever possible to reduce the attenuation in the media. The thickness mode resonant frequencies of commercially available piezoelectric films are usually between a few MHz and a few tens of MHz. Of course the resonant frequencies will be affected by the attached structure, size, shape and boundary conditions of the sensor's active element. Thick, low frequency films are recommended since a few MHz should be high enough for most flow imaging purposes. Also in this frequency region the effective thickness of the pipe wall will be a few times the wave length. Therefore the fundamental guided wave phenomena, which would result in a stronger interfering effect, will not occur.

In fact, even if some of the direct transmissions through the pipe are

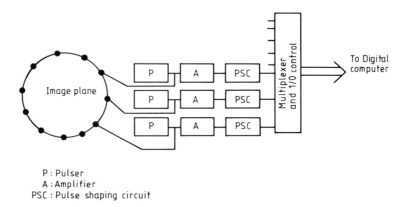

P : Pulser
A : Amplifier
PSC : Pulse shaping circuit

Figure 3.32 Block diagram of electric circuits in an ultrasonic imaging system (only three channels are shown.

noticeable, they need not corrupt the proper measurements because they are not random and can be eliminated by timing and amplitude discrimination. In most cases timing alone will be sufficient (Chapter 4.6).

For both reflection and transmission modes, the acoustic energy passes at least twice through the boundaries between the pipe and the liquid, from the transmitting lens to the flow and from the flow to the receiving lens. If possible, the material of the pipe should be chosen so that its acoustic impedance is as close as possible to that of the liquid. This will result in the most effective energy transmission and least reflective losses (Lynnworth 1965). Unfortunately, in most practical cases, a close match is not possible.

Electronic system considerations. The electronic system required to control the ultrasonic imaging sensor has four main functions:

1. Supplying pulses to activate the transmitting sensors. These pulses should ideally be software controlled so that the timing of the pulses can be easily varied and synchronization ensured.
2. Amplifying the analogue signals from the receiving sensors.
3. Reshaping the received analogue signals into digital pulses which preserve the time of arrival information.
4. Interfacing to the digital computer for control of pulse generation and image reconstruction from received data.

A functional block diagram of the electronic system is shown in figure 3.32.

3.7 SUMMARY

- A wide variety of sensing techniques can be used for flow imaging.
- The best sensing technique for a particular application will depend upon the

actual components to be imaged, the flow velocity, the resolution required and the environment in which the measurement is to be made.

- In view of the relatively high development costs of making practical flow imaging systems the use of CAD techniques to develop concepts is to be encouraged.

4

Image reconstruction

4.1 INTRODUCTION

Computerized tomography is the computer-assisted reconstruction of the cross sectional image of an object on the basis of measurements of signals that have passed through that object. In flow imaging the objective is usually to determine the cross-sectional density distribution of a multi-component flow.

Image reconstruction techniques have been extensively developed for medical and industrial applications; however, two particular problems are encountered with flow imaging:

1. The flow pattern changes rapidly, so efficient data processing at an acceptable cost is required.

2. The field pattern of certain sensors may vary with the distribution of the components in the field.

The latter problem is encountered especially with electrical capacitance and resistance field sensing methods (Chapter 3). Interaction between the sensing field and the flow components (flow field) is unlikely to be significant with ultrasound systems, although scattering could introduce some difficulties. It will not generally be a problem with radiation attenuation systems such as gamma and optical attenuation where the measured discontinuities in the flow are large compared with the wavelength of the radiation.

4.2 BASIC PRINCIPLES OF IMAGE RECONSTRUCTION

In this section the basic principles of image reconstruction will be examined using the example of a gamma radiation measurement system.

In gamma tomography the cross section of an object is scanned by a parallel or divergent gamma beam, figure 3.2(a,b). Its attenuation is recorded by a detector or an array of detectors and then processed by a computer to obtain the spatial distribution of the attenuation coefficients. Referring to figure 4.1, the gamma attenuation dI from point z to $z + dz$ is

$$dI/I = \exp(-\mu(z)\,dz) \tag{4.1}$$

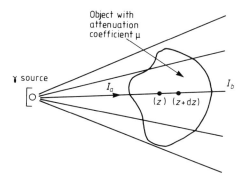

Figure 4.1   Showing attenuation of γ-ray along line z.

where I is the γ-ray intensity and $\mu(z)$ the linear absorption coefficient. When passing through the object along the straight line L from point a to point b, the γ-ray is attenuated as follows

$$\frac{I_a}{I_b} = \exp\left(-\int_{L(a)}^{L(b)} \mu(z)\,dz\right).$$
(4.2)

The value of the line integral (4.2) is called the *ray sum*. The scanning process provides a set of such ray sums. The set is known as a *projection*. The aim of the reconstruction process is to reconstruct the spatial attenuation function of the object, $\mu(x, y)$ in figure 4.1 from these integrals, i.e. to reconstruct the two-dimensional distribution of the attenuation coefficient in terms of the co-ordinate system (x, y). In order to do this intersecting projections from many angles are required (figure 3.1(b)).

Let J be the image space (the space of real functions on the two-dimensional Euclidean space) and P the projection space (the space of the line integral on J). In figure 4.2 ℓ is the distance of the line L from the co-ordinate system origin, Θ is the angle between L and the positive y axis, and z is the distance along line L from point $z = 0$. The transform which is defined by

$$Rf(\ell, \theta) = \int_{-\infty}^{\infty} f(\ell\cos\theta - z\sin\theta, \ell\sin\theta + z\cos\theta)\,dz$$
(4.3)

is called the two-dimensional Radon transform. The function under the integral represents the attenuation coefficient of the object $f(x, y) = \mu(x, y)$. Hence reconstruction of the object requires the inverse Radon transformation R^{-1}. The inverse Radon transform was given by Radon (1917), but the formula is of limited value from a practical point of view because in flow imaging one deals

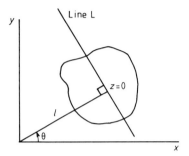

Figure 4.2   The two-dimensional Radon transform.

with finite and imperfect measurement data. The reconstruction problem then is to estimate the values of the function at given points from a partial knowledge of its Radon transformation, in other words, to develop techniques for inverting the integral (4.3). No single technique has yet been found capable of satisfactorily processing the wide variety of projection-measurement geometries which occur in practical applications.

To choose or develop a suitable algorithm for a particular application is by no means a trivial matter. Although many different problems may have the same underlying mathematical description, specific solutions may differ for each application. One reason for this is that the measurement sensors and the requirements for the image vary from application to application. Generally speaking, a good algorithm should be fast and accurate, need minimum computing, and be easy to understand.

In order to define the inverse problem of image reconstruction, it is first necessary to define the direct (or forward) problem. The direct problem is that of mapping a set of theoretical 'parameters' into a set of experimentally measured 'results' (Radon 1917, Herman 1980). Finding the 'parameters' from given measured 'results' is called solving the inverse problem. The first step in solving an inverse problem is to solve the direct problem, otherwise no solution of the inverse problem can be found. Once this is done, the direct mapping can be constructed.

The direct mapping problem can be defined as:

$$x \rightarrow y \tag{4.4}$$

where x is a vector representing the component distribution in a two-component flow

$$x = [x_1 x_2 \ldots x_i \ldots x_N] \tag{4.5}$$

where the x_i is the component fraction in the ith cell (pixel) of the image space. N is the number of cells in the mesh and y is a vector constructed from the known measurements.

$$y = [y_1 y_2 \ldots y_i \ldots y_M] \tag{4.6}$$

where y_i is the ith measurement and M is the number of available measurements.

Obviously one cannot construct the mapping until the nature of the 'results' is precisely defined. A simple expression is desirable in order to assist the construction of its inverse. With the mapping being presented by a matrix relationship (4.4) can be rewritten as:

$$y = \mathbf{R}x \tag{4.7}$$

where $\mathbf{R}$ is an $(M \times N)$ matrix that represents the direct mapping of the problem and is called the projection matrix. In practical applications, the vector y is only known approximately since it is obtained from measurements. In such cases, it is only possible to find an approximate solution to the equation

$$\mathbf{R}x = \tilde{y} \tag{4.8}$$

where $\tilde{y}$ is an approximation of the vector y. However, this approximate solution

$$x = \mathbf{R}^{-1}\tilde{y} \tag{4.9}$$

is not usually valid for two reasons:

1. such a solution may not exist in the set of solutions that contains all physically possible component fractions.

2. even if such a solution does exist, it may not possess the property of stability, that is, a small change in $\tilde{y}$ may result in a large change in x.

However, with the aid of supplementary information on the nature of the solution, one can obtain a quasi-solution by minimizing some function, an example of which is a 'least-squares' solution. In its simplest form the problem may be regarded as one of finding the solution to a set of linear equations:

$$\underset{M \times 1}{\tilde{y}} = \underset{M \times N}{\mathbf{R}} \underset{N \times 1}{x} \tag{4.10}$$

where the vector $\tilde{y}$ contains data measured externally by sensors, the vector x is a set of unknown coefficients characterising the instantaneous cross section, and the matrix $\mathbf{R}$ contains known geometric factors. For example, if the sensors measure the local attenuation of radiation by pixels or picture elements throughout the cross section, y contains 'projection' data indicating the total attenuation of beams along straight lines through the cross section, and the matrix $\mathbf{R}$ encodes the contribution that any pixel makes towards the observed total attenuation of the beam y.

In imaging a flow, two limitations are apparent in the above formulation, both arising because of the short time window available for taking measurements:

1. The number of measurements M relating to a single cross section is limited, which places a restriction on the number of variables N. This imposes a limit on the spatial resolution achievable for flow imaging systems. Thus, unlike medical tomographic imaging systems in which reconstruction is performed from an effectively 'complete' set of projections taken at say $1°$ angular intervals, flow imaging has to contend with a dramatically reduced set of projections, possibly as few as ten, sometimes spaced at irregular angular intervals.

2. The signal-to-noise ratio in the measured values y is often low.

Statistical approaches to what has classically been regarded as an inversion problem have been proposed in the literature principally for the analysis of data from emission tomography. In the so-called missing data technique the problem of determining the set of attenuation coefficients, x, is regarded as one of statistical estimation rather than numerical inversion. In these circumstances the principle of maximum likelihood can be used and the task of estimation is then founded on well developed statistical theory. A further advantage of this approach is the possibility of obtaining improved estimates of x (i.e. better visual images of flow cross sections) by applying techniques for determining smoothing estimators. Smoothing of the expected flow region can be regarded as a way of incorporating *a priori* knowledge into the reconstruction process. For the imaging of multi-component flows which do not mix, it is possible to incorporate into the reconstruction algorithm the knowledge that the attenuation coefficients being determined can only take a discrete number of levels corresponding to the individual components, water, oil, gas etc. In principle, the uses of such prior knowledge should dramatically reduce the data requirements for acceptable reconstruction. This simple reconstruction method, suitable for a restricted class of flow imaging systems, will be described in section 4.3. The subsequent sections 4.4, 4.5 and 4.6 will describe the special reconstruction methods used for capacitive impedance, resistive impedance and ultrasonic flow tomography.

4.3   TWO-COMPONENT FLOW IMAGE RECONSTRUCTION USING A TWO-PROJECTION KNOWLEDGE-BASED RECONSTRUCTION ALGORITHM

The reconstruction algorithms described in later sections of this chapter were designed to reconstruct images where the cross- section may have a wide range of densities, contrasts, etc. A few industrial flows contain only two components in well separated regions (oil/water, liquid/gas, gas/solids etc). The two-projection reconstruction algorithm (developed by the authors) which will now be described, is specifically designed to deal with the two-component situation. It has a specific advantage of computational simplicity and it requires a simple sensor arrangement, but will only generate images where each pixel is

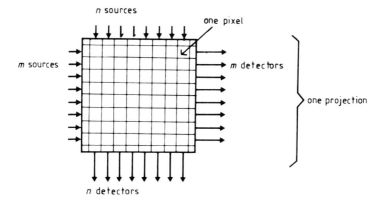

Figure 4.3   Source and detector arrangement for two projection systems.

completely filled with one or other of the two components. Any pixel containing a mixture would be imaged to show the pixel to be occupied by the major component.

The principal features of the method are:

1. Two orthogonal projections are used (figure 4.3). There are m parallel beams in one projection and n parallel beams in the other. Each beam is regarded as having one source and one detector.

2. The output signals from the detectors are preprocessed so that we obtain a zero value if the beam traverses only the component with the lower attenuation coefficient, and a maximum value if the beam traverses only the component with the higher attenuation coefficient. Between these limits the processed detector signal is linear with respect to component concentration ratio. It is convenient to allow the maximum value of the processed detector output to be equal to the number of pixels along the direction of the ray. Thus if we consider a system with nine sources and detectors in each direction ($m = n = 9$), the process signals from each projection are in the range zero to 9. (Note it is usual to use an odd number of rays in each projection so that one pixel exists in the centre of the image—this enables any interesting feature in the centre of the flow to be imaged).

3. Any non-rectangular pipe geometries are transformed (remapped) into a rectangular array.

The procedure for reconstruction will now be explained using the following notation:

• the first projection (measurement) signal (ray sum) is

$$X = [X_1, X_2, \ldots, X_m] \tag{4.11}$$

• the second projection (measurement) signal (ray sum) is

$$Y = [Y_1, Y_2, \ldots, Y_n] \tag{4.12}$$

- the image matrix grey level is:

$$G(n, m) = \begin{bmatrix} g_{11} & \cdots & g_{1n} \\ & \cdots & \\ g_{m1} & \cdots & g_{mn} \end{bmatrix} \qquad (4.13)$$

- the pixel value g_{ij} is binary, i.e. $g_{ij} = 0$ if the first component A occupies the pixel, and $g_{ij} = 1$ if the second component B occupies the pixel;
- the number of pixels in the image matrix is $N = mn$. In our example $N = 9 \times 9 = 81$;
- the fraction of the second component α_B in the cross-section is

$$\alpha_B = \frac{N_B}{N} \times 100\% \qquad (4.14)$$

where N_B is the number of pixels containing component B

$$N_B = \sum_1^m X_m = \sum_1^n Y_n.$$

The measured projection values (y in (4.4)) are a set of the measured values of the first projection (4.11) and a set of the measured values of the second projection (4.12). The geometric factors of matrix **R** in (4.10) are defined as the linear ray sum projections:

$$X_j = \sum_{r=1}^m w_{rj} g_{rj}$$

$$\qquad (4.15)$$

$$Y_i = \sum_{r=1}^n w_{ri} g_{ri}$$

where

$$\begin{cases} g_r = 1 & \text{if only the second component } B \text{ is present in the pixel} \\ 0 & \text{if only the first component } A \text{ is present in the pixel} \end{cases}$$

and w are the measurement detector sensitivity factors corresponding to each pixel.

Example of the reconstruction method. Considering the measured ray sums (equations (4.11) and (4.12)) we can see how the distribution of the second component could be

$$X = [0, 0, 3, 4, 9, 0, 0, 0, 0] \qquad \text{and}$$
$$Y = [2, 3, 3, 3, 1, 1, 1, 1, 1]$$

(these correspond to X_j and Y_i in equation (4.15). The measurement matrix is:

$$
\begin{pmatrix}
\cdot & \cdot & \cdot & \cdot & \cdot & \cdot & \cdot & \cdot & \cdot & Y1 = 2 \\
\cdot & \cdot & \cdot & \cdot & \cdot & \cdot & \cdot & \cdot & \cdot & Y2 = 3 \\
\cdot & \cdot & \cdot & \cdot & \cdot & \cdot & \cdot & \cdot & \cdot & Y3 = 3 \\
\cdot & \cdot & \cdot & \cdot & \cdot & \cdot & \cdot & \cdot & \cdot & Y4 = 3 \\
\cdot & \cdot & \cdot & \cdot & \cdot & \cdot & \cdot & \cdot & \cdot & Y5 = 1 \\
\cdot & \cdot & \cdot & \cdot & \cdot & \cdot & \cdot & \cdot & \cdot & Y6 = 1 \\
\cdot & \cdot & \cdot & \cdot & \cdot & \cdot & \cdot & \cdot & \cdot & Y7 = 1 \\
\cdot & \cdot & \cdot & \cdot & \cdot & \cdot & \cdot & \cdot & \cdot & Y8 = 1 \\
\cdot & \cdot & \cdot & \cdot & \cdot & \cdot & \cdot & \cdot & \cdot & Y9 = 1 \\
X1 & X2 & X3 & X4 & X5 & X6 & X7 & X8 & X9 & \\
=0 & 0 & 3 & 4 & 9 & 0 & 0 & 0 & 0 &
\end{pmatrix}
$$

Thus the value of the ray sums is known but not which pixels are equal to 1 and which are zero. However, since $Y1 = 2$ then there are two pixels equal to 1 along ray $Y1$. Thus the number of pixels in a cross-section equal to 1 is the sum of the ray sums of X_j or Y_i. Thus the number of pixels (N_B) occupied by component B having value $= 1$ is

$$
N_B = \sum_{j=1}^{9} X_j = \sum_{i=1}^{9} Y_i = 16
$$

and the ratio of component B in the cross-section ((4.14) is therefore

$$
\alpha_B = (16/81)x\,100\% = 19.8\%.
$$

If $X_j = 0$ or $Y_i = 0$ it means that all pixels where these rays intersect are occupied by component A so $g_{ij} = 0$ (this information is stored). For the next step we consider the following matrices, the first where zeros are inserted for appropriate pixels, and secondly derived from this a reduced 9×3 matrix of all non-zero rows and columns:

$$
\begin{pmatrix}
0 & 0 & \cdot & \cdot & \cdot & 0 & 0 & 0 & 0 & & \cdot & \cdot & \cdot & Y1 = 2 \\
0 & 0 & \cdot & \cdot & \cdot & 0 & 0 & 0 & 0 & & \cdot & \cdot & \cdot & Y2 = 3 \\
0 & 0 & \cdot & \cdot & \cdot & 0 & 0 & 0 & 0 & & \cdot & \cdot & \cdot & Y3 = 3 \\
0 & 0 & \cdot & \cdot & \cdot & 0 & 0 & 0 & 0 & \Rightarrow & \cdot & \cdot & \cdot & Y4 = 3 \\
0 & 0 & \cdot & \cdot & \cdot & 0 & 0 & 0 & 0 & & \cdot & \cdot & \cdot & Y5 = 1 \\
0 & 0 & \cdot & \cdot & \cdot & 0 & 0 & 0 & 0 & & \cdot & \cdot & \cdot & Y6 = 1 \\
0 & 0 & \cdot & \cdot & \cdot & 0 & 0 & 0 & 0 & & \cdot & \cdot & \cdot & Y7 = 1 \\
0 & 0 & \cdot & \cdot & \cdot & 0 & 0 & 0 & 0 & & \cdot & \cdot & \cdot & Y8 = 1 \\
& & & & & & X3 & X4 & X5 & & & & & \\
& & & & & = & 3 & 4 & 9 & & & & &
\end{pmatrix}
$$

The next step is to find rows or columns with all pixels occupied by component B, i.e. with the value of the ray sum equal to the number of pixels. In this example $X5 = 9$, $Y2 = 3$, and $Y3 = 3$, and $Y4 = 3$. If pixel values of unity are inserted in all these rows and columns then a matrix having only unknown pixel values is obtained.

$$
\begin{pmatrix}
0 & 0 & . & . & 1 & 0 & 0 & 0 & 0 \\
0 & 0 & 1 & 1 & 1 & 0 & 0 & 0 & 0 \\
0 & 0 & 1 & 1 & 1 & 0 & 0 & 0 & 0 \\
0 & 0 & 1 & 1 & 1 & 0 & 0 & 0 & 0 \\
0 & 0 & 0 & 0 & 1 & 0 & 0 & 0 & 0 \\
0 & 0 & 0 & 0 & 1 & 0 & 0 & 0 & 0 \\
0 & 0 & 0 & 0 & 1 & 0 & 0 & 0 & 0 \\
0 & 0 & 0 & 0 & 1 & 0 & 0 & 0 & 0 \\
0 & 0 & 0 & 0 & 1 & 0 & 0 & 0 & 0
\end{pmatrix}
\Rightarrow
\begin{matrix}
 & . & & . & Y1 = 1 \\
 & & & & \\
 & x_3 & & x_4 & \\
 & & = 0, & 1 &
\end{matrix}
$$

Following the procedure described above, the final 1×2 matrix is solved as being [0 1], so the reconstructed image is:

$$
\begin{pmatrix}
0 & 0 & 0 & 1 & 1 & 0 & 0 & 0 & 0 \\
0 & 0 & 1 & 1 & 1 & 0 & 0 & 0 & 0 \\
0 & 0 & 1 & 1 & 1 & 0 & 0 & 0 & 0 \\
0 & 0 & 1 & 1 & 1 & 0 & 0 & 0 & 0 \\
0 & 0 & 0 & 0 & 1 & 0 & 0 & 0 & 0 \\
0 & 0 & 0 & 0 & 1 & 0 & 0 & 0 & 0 \\
0 & 0 & 0 & 0 & 1 & 0 & 0 & 0 & 0 \\
0 & 0 & 0 & 0 & 1 & 0 & 0 & 0 & 0 \\
0 & 0 & 0 & 0 & 1 & 0 & 0 & 0 & 0
\end{pmatrix}.
$$

Modified approach to resolve ambiguous cases.   Unfortunately the above procedure is not satisfactory in all cases. For example, if single component B pixels occur in each row and column we cannot determine their positions by the method above, i.e. if the image is:

$$
\begin{pmatrix}
1 & 0 & 0 & 0 & 0 & 0 & 0 & 0 & 0 \\
0 & 1 & 0 & 0 & 0 & 0 & 0 & 0 & 0 \\
0 & 0 & 1 & 0 & 0 & 0 & 0 & 0 & 0 \\
0 & 0 & 0 & 1 & 0 & 0 & 0 & 0 & 0 \\
0 & 0 & 0 & 0 & 1 & 0 & 0 & 0 & 0 \\
0 & 0 & 0 & 0 & 0 & 1 & 0 & 0 & 0 \\
0 & 0 & 0 & 0 & 0 & 0 & 1 & 0 & 0 \\
0 & 0 & 0 & 0 & 0 & 0 & 0 & 1 & 0 \\
0 & 0 & 0 & 0 & 0 & 0 & 0 & 0 & 1
\end{pmatrix}
\quad
\begin{aligned}
X &= [1, 1, 1, 1, 1, 1, 1, 1, 1] \\
Y &= [1, 1, 1, 1, 1, 1, 1, 1, 1]
\end{aligned}.
$$

Similarly, if the image is

$$\begin{pmatrix} 0 & 0 & 0 & 0 & 1 & 0 & 0 & 0 & 0 \\ 0 & 0 & 1 & 0 & 0 & 0 & 0 & 0 & 0 \\ 0 & 0 & 0 & 0 & 0 & 0 & 0 & 1 & 0 \\ 0 & 1 & 0 & 0 & 0 & 0 & 0 & 0 & 0 \\ 0 & 0 & 0 & 0 & 0 & 1 & 0 & 0 & 0 \\ 0 & 0 & 0 & 0 & 0 & 0 & 0 & 0 & 1 \\ 1 & 0 & 0 & 0 & 0 & 0 & 0 & 0 & 0 \\ 0 & 0 & 0 & 1 & 0 & 0 & 0 & 0 & 0 \\ 0 & 0 & 0 & 0 & 0 & 0 & 1 & 0 & 0 \end{pmatrix}$$

$$X = [1, 1, 1, 1, 1, 1, 1, 1, 1]$$
$$Y = [1, 1, 1, 1, 1, 1, 1, 1, 1].$$

In both the above cases the measured row and column vectors are identical, although the images are obviously different.

In both cases the fraction of B in the cross-section would be correctly estimated at $\alpha_B = 11\%$. However, the problem of ambiguous pixel positions arises; in other words, there are fewer equations than unknown values if the procedure described above is followed. In such cases, provided that the geometrical regime of the flow pattern is known, this difficultly can be reduced by using an iterative search and shrink routine which enables the elements in the image matrix to be efficiently determined. Results for stratified and annular flow patterns using this iterative routine are given in figure 4.4 and figure 4.5. The appropriate flow regime is assumed to give projection data X and Y as shown. From this projection data the images are reconstructed using the flowchart shown in figure 4.6. The reconstructed images show a reasonable agreement within the limitations imposed by using only two projections.

4.4 RECONSTRUCTION ALGORITHM—CAPACITIVE IMPEDANCE SENSING TECHNIQUE

This reconstruction algorithm is designed for the capacitance sensing technique described in Chapter 3.4 (Huang *et al* 1989). The algorithm was developed for two-phase flow, and a sensor with eight electrodes giving 28 independent measurements was used.

The vector of the measurement results $[m_1, m_2, \ldots m_M]$ (actually the difference between the capacitance with the material present and the standing capacitance, sections 3.4.3 and 3.4.7.2) is determined by the distribution of dielectric material in the pipe cross section $\epsilon(x, y)$, i.e.

$$[m_1, m_2, \ldots m_M] = f(\epsilon(x, y)). \tag{4.16}$$

The task of image reconstruction is to solve the inverse problem of determining $\epsilon(x, y)$ from this limited number of measurements. Since the component distribution in two- component flows is complex, with only 28 measurements a complete solution of this problem is not possible. Therefore a linear approximation approach known as linear backprojection is used.

Projection   Model   Reconstruction

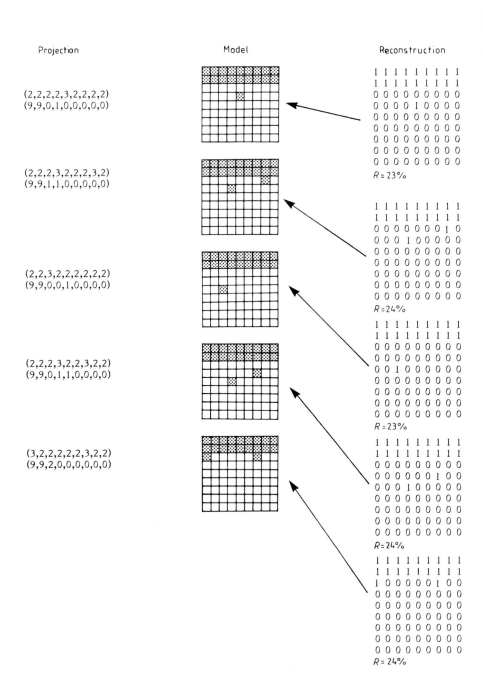

(2,2,2,2,3,2,2,2,2)
(9,9,0,1,0,0,0,0,0)

(2,2,2,3,2,2,2,3,2)
(9,9,1,1,0,0,0,0,0)

(2,2,3,2,2,2,2,2,2)
(9,9,0,0,1,0,0,0,0)

(2,2,2,3,2,2,3,2,2)
(9,9,0,1,1,0,0,0,0)

(3,2,2,2,2,2,3,2,2)
(9,9,2,0,0,0,0,0,0)

Figure 4.4   Stratified flow.

The simplicity of linear back projection is illustrated in figure 4.7. An arrangement using only three pairs of electrodes in the X and Y direction is shown, The shaded areas represent a fluid or solid in the vessel which is electrically different from the material around it. If, as in cases 1 and 2, we look at the X and Y directions, it is obvious where the areas of material are, because of the X and Y overlap. But, in the third case the method throws up an inconsistency. This is why it is necessary to look at a system in more than two directions (e.g. the 28 directions shown in figure 4.9 below) otherwise the results can be misleading. To simplify the reconstruction problem it is assumed that a capacitance measurement results from a homogenous change in the permittivity over the entire positive sensing area of the appropriate electrode pair (figure 4.8).

Each positive sensing area is given a grey level whose value is proportional to the measured capacitance. By superimposing the grey levels of the 28 sensing areas (figure 4.9), the grey level in the area where the object is present will be enhanced. The basic principles of the linear backprojection algorithm used in this system will now be considered. The first step in reconstructing the image is to find the positive sensing areas of all 28 measurements (Chapter 3.4). An input data file to the main reconstruction program is then created containing information on the geometric shape, size, and position of the image segments formed by the 28 positive sensing areas. Each of the segments is related uniquely to one or more of the 28 sensing areas, i.e. is inside these areas. Therefore each segment can be related to a 28-element vector V, where each element has the value 1 if the segment is inside the corresponding sensitive areas, otherwise it is zero. The grey level G of each segment can be obtained by multiplying the corresponding segment vector V by the 28 measurement values $[m_1 \dots m_{28}]$ Therefore the array of the image segments is related to the vector of the measured data by

$$
\begin{vmatrix} G1 \\ G2 \\ ,, \\ ,, \\ ,, \\ ,, \\ Gk \end{vmatrix} = \begin{vmatrix} V1 \\ V2 \\ ,, \\ ,, \\ ,, \\ ,, \\ Vk \end{vmatrix} \begin{vmatrix} m_1 \\ m_2 \\ ,, \\ ,, \\ ,, \\ ,, \\ m_{28} \end{vmatrix} \tag{4.17}
$$

where k is the total number of the segments. In this case $k = 363$. Equation (4.17) shows that the image reconstruction problem has been simplified to a linear problem of matrix multiplication. The 363-row by 28-column matrix, [**V**], is produced by examining the relationship between each of the 363 segments and the 28 measurements. Each of the 363 rows represents a segment in figure 4.9 while each of the 28 columns corresponds to one of the 28 measurements. Since the value of any element in the matrix is either one or zero (i.e. a specific segment is either inside or outside the sensing zones of the sensor), one element can be expressed by a one-bit digit or each row of the 28-elements can be

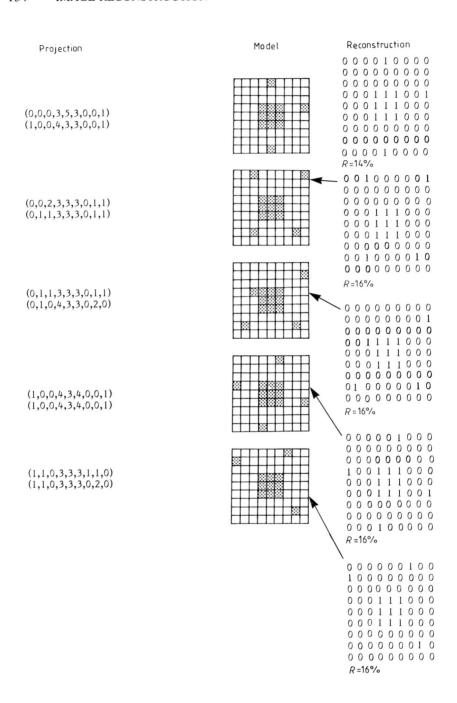

Projection

Model

Reconstruction

(0,0,0,3,5,3,0,0,1)
(1,0,0,4,3,3,0,0,1)

```
0 0 0 0 1 0 0 0 0
0 0 0 0 0 0 0 0 0
0 0 0 0 0 0 0 0 0
0 0 0 1 1 1 0 0 1
0 0 0 1 1 1 0 0 0
0 0 0 1 1 1 0 0 0
0 0 0 0 0 0 0 0 0
0 0 0 0 0 0 0 0 0
0 0 0 0 1 0 0 0 0
```
R = 14%

(0,0,2,3,3,3,0,1,1)
(0,1,1,3,3,3,0,1,1)

```
0 0 1 0 0 0 0 0 1
0 0 0 0 0 0 0 0 0
0 0 0 0 0 0 0 0 0
0 0 0 1 1 1 0 0 0
0 0 0 1 1 1 0 0 0
0 0 0 1 1 1 0 0 0
0 0 0 0 0 0 0 0 0
0 0 1 0 0 0 0 1 0
0 0 0 0 0 0 0 0 0
```
R = 16%

(0,1,1,3,3,3,0,1,1)
(0,1,0,4,3,3,0,2,0)

```
0 0 0 0 0 0 0 0 0
0 0 0 0 0 0 0 0 1
0 0 0 0 0 0 0 0 0
0 0 1 1 1 1 0 0 0
0 0 0 1 1 1 0 0 0
0 0 0 1 1 1 0 0 0
0 0 0 0 0 0 0 0 0
0 1 0 0 0 0 0 1 0
0 0 0 0 0 0 0 0 0
```
R = 16%

(1,0,0,4,3,4,0,0,1)
(1,0,0,4,3,4,0,0,1)

```
0 0 0 0 0 1 0 0 0
0 0 0 0 0 0 0 0 0
0 0 0 0 0 0 0 0 0
1 0 0 1 1 1 0 0 0
0 0 0 1 1 1 0 0 0
0 0 0 1 1 1 0 0 1
0 0 0 0 0 0 0 0 0
0 0 0 0 0 0 0 0 0
0 0 0 1 0 0 0 0 0
```
R = 16%

(1,1,0,3,3,3,1,1,0)
(1,1,0,3,3,3,0,2,0)

```
0 0 0 0 0 0 1 0 0
1 0 0 0 0 0 0 0 0
0 0 0 0 0 0 0 0 0
0 0 0 1 1 1 0 0 0
0 0 0 1 1 1 0 0 0
0 0 0 1 1 1 0 0 0
0 0 0 0 0 0 0 0 0
0 0 0 0 0 0 0 1 0
0 0 0 0 0 0 0 0 0
```
R = 16%

Figure 4.5   Annular flow.

expressed using one long integer rather than using 28 integers, in order to save memory space.

So the grey level of the 363 segments is calculated by multiplying the matrix [**V**] by vector m of the 28 measured values. The grey level of each segment is calculated by testing each bit of the corresponding row in [**V**], summing all measured values which correspond to the 1 bits and ignoring those corresponding to the 0 bits. The sum is divided by the total number of the 1 bits in this row.

Weighting and threshold filtering are used to improve the image (Chapter 5). Since the measurement sensitivity in the central areas (Chapter 3.4) of the pipe is considerably lower than it is in the areas near the pipe wall, the grey level of the segment pixels in the central area is weighted according to the function

$$\tilde{G}_2(r) = [\tilde{G}_1(r)]W(r) \qquad (4.18)$$

where $\tilde{G}_1$ and $\tilde{G}_2$ are the pixel grey levels before and after weighting and $W(r)$ is the weighting factor at radius r from the centre of the pipe.

After the above steps, the 363 grey levels are obtained. The 363 segments are then displayed using these grey levels.

The reconstruction program consists mainly of a single matrix multiplication. The most time consuming task in the software design is to produce the 28×363 matrix defining the 28 positive sensing areas. Once this matrix has been produced, it will be satisfactory unless there are large changes in the permittivity ratios of the components being imaged. However, when two components with a very large permittivity ratio, such as water and gas, are being imaged, the distribution of the positive sensing areas will be distorted (Chapter 3.2.2).

4.5 RECONSTRUCTION ALGORITHMS–RESISTIVE IMPEDANCE SENSING TECHNIQUE

The instrumentation used for impedance imaging conducting fluids was described in Chapter 3.5 and is shown in schematic block- diagram form in figure 3.25. It is composed of three parts: the electrodes connected (non-intrusively) to the vessel, the data acquisition system (DAS), and the image reconstruction system (IRS). Electrodes are placed into the pipe wall in order to make measurements of the distribution of electrical resistance within an image plane. This is performed by injecting an ac current via one pair of adjacent electrodes and measuring the voltage across all other pairs of adjacent electrodes. The procedure is repeated for all possible combinations of driving pair electrodes as shown in figure 3.25. The number of unique measurements M (allowing for reciprocity) obtained using this protocol with N electrodes is determined by the following relationship:

$$M = \frac{N(N-3)}{2}. \qquad (4.19)$$

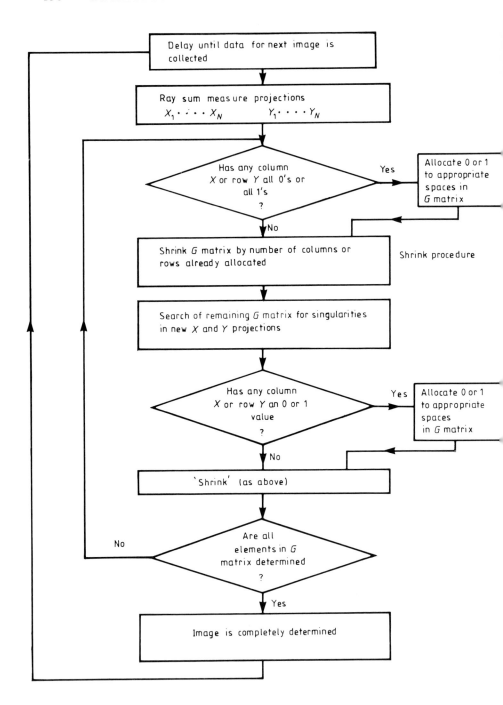

Figure 4.6   Search-and-shrink routine for two-projection flow imaging.

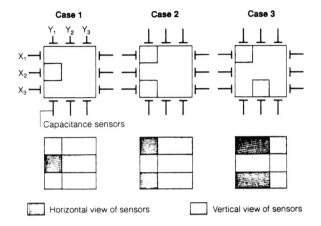

Figure 4.7   Back projection method.

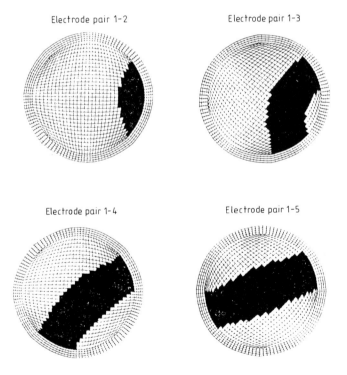

Figure 4.8   The positive sensing area of electrode pairs 1–2, 1–3, 1–4 and 1–5.

Four electrodes are used in this technique to ensure that any variation in contact impedance between the current injecting electrodes and the material inside the pipe does not influence the measured voltage. Conversely, in situations

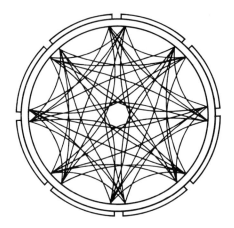

Figure 4.9 The image segments produced by the intercepting positive sensing areas.

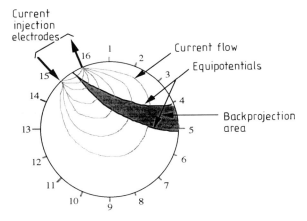

Figure 4.10 Showing backprojection of voltage difference between equipotential lines.

where the contact impedance is negligible, N additional measurements may be obtained by reverting to the 2-electrode technique whereby voltage readings can be obtained from all electrode pairs including the current driving pair. Thus, the number of unique measurements for the 2-electrode technique is given by

$$M = \frac{N(N-1)}{2}. \tag{4.20}$$

In this section it will be assumed that the 4-electrode technique is used.

The reconstruction algorithm can be thought of simply as a series of procedures performed repeatedly on digitized measurement data from the imaging system electronics in order to determine the distribution of regions of different resistivities (e.g. component concentrations) within a cross section of the pipe.

To appreciate the demands placed on the reconstruction algorithm it is first necessary to describe the nature of the reconstruction problem. Initially, consider the two-dimensional circular cross section in figure 4.10 which has a number of equally-spaced electrodes on the periphery. Given that the material inside the region has resistivity ρ and that the applied boundary currents are also known, then the voltages at any point inside the pipe or on the periphery are given by

$$\nabla \rho^{-1} \nabla V = 0 \qquad (4.21)$$

with the following boundary condition

$$\rho^{-1} \frac{\delta V}{\delta n} = J_0$$

where ρ is the resistivity at any point inside the region, $\delta V / \delta n$ is the normal derivative to the surface, and J_0 is the normal component of the electric current density at the surface.

Equation (4.21) is Poisson's equation and the solution of the partial differential equation is referred to as the solution of the *forward* problem— forward because ρ is known everywhere inside the vessel (i.e. the task is to find V inside and on the periphery of the vessel, given J_0 and ρ at all points). The situation is that a limited number of voltage measurements are obtained in response to the constant currents applied to the electrodes on the periphery of the pipe. From these measurements we must reconstruct an accurate representation of the resistivity profile inside the pipe to solve the nonlinear inverse problem (i.e. the task is to find ρ given J_0 and V on the periphery). In mathematical terms this is referred to as a boundary value problem with Neumann boundary conditions.

There are a number of methods for solving the forward problem based on experimental, analytical or numerical solutions. The classical experimental method of solution involves the use of physical models placed into electrolytic tanks or painted onto teledeltos paper (Liebmann 1953). Analytical solutions are generally based on series expansion methods which can be further simplified if the pipe has axes of symmetry and a regular boundary profile.

The numerical solution approach is more popular since there are many numerical subroutine libraries available for computers. The finite element method (FEM) approximation is a favoured approach which discretizes the region of interest into a mesh-like structure of triangular or quadrilateral elements. The linear system of equations describing the state of each element under given boundary conditions is then solved algebraically using appropriate computer software subroutines.

The solution of the inverse problem can be performed numerically using direct or indirect techniques resulting in linear approximations of the nonlinear problem. The reconstruction time determines whether the measured data can be displayed in real time. Workers concerned with medical uses of impedance imaging have developed a number of interesting reconstruction algorithms, the most promising of which have been analysed recently by Yorkey (1986, 1990) at the University of Wisconsin. For example, one of the fastest reconstruction algorithms was the backprojection algorithm developed at Sheffield University, UK, for their 16-electrode applied potential tomography (APT) system which was the world's first commercially available clinical impedance tomography instrument (Barber and Brown 1984). At the opposite end of the reconstruction-complexity spectrum is the algorithm devised by Yorkey (1990) which is accepted as being one of the most theoretically correct reconstruction algorithms. Unfortunately, the consequence of this 'correctness' is that the algorithm is computationally intensive and therefore less practical to implement, although recent developments on this are more efficient and provide good quantitative images (Abdullah *et al* 1992). The following sub-sections briefly describe the salient features of the backprojection algorithm and of Yorkey's algorithm.

4.5.1 The backprojection method

The technique used in this algorithm is referred to as backprojection between equi-potential lines since the potential difference on the surface (i.e. from adjacent electrodes) is backprojected to a resistivity value in the area between the two equipotential lines, as shown in figure 4.10, namely

$$\rho_{\text{calculated}} = \rho_{\text{initial guess}} \frac{(V_D)_{\text{measured}}}{(V_D)_{\text{expected}}} \tag{4.22}$$

where V_D is the potential difference between two equipotential lines. Initially, the region between the equipotential lines is assumed to have homogeneous resistivity for which the potential differences at the electrodes are calculated using finite element or other suitable techniques. To obtain a backprojection image the values of ρ are averaged over the entire image after equation (4.22) has been solved for all projections. In order to improve the quality the averaged image can then be filtered by a ramp filter to reduce blurring (Seagar *et al* 1987). Additional operations can then be performed on each of the image pixels in order to accommodate for the fact that the equipotential field lines are nonlinear and also to minimize errors due to the severe position dependence of the spatial resolution of the reconstructed image.

4.5.2 Newton–Raphson reconstruction algorithm

This algorithm, devised by Yorkey, is so-called because it makes extensive use of the Newton-Raphson iterative minimization algorithm to solve the inverse problem (Yorkey 1990). The overall function of the reconstruction algorithm can be summarized by the following segment of meta-code:

Guess initial resistivity distribution for whole region
REPEAT
    Enter resistivity values into finite element model
    Solve forward problem
    Compare solution of forward problem with measured values to obtain an
    error value
    Update resistivities
UNTIL (error value $\leqslant$ predefined limit)

The performance of the Newton–Raphson algorithm in iterating to a low error value depends largely on the initial guess of the resistivity distribution. If the guess is poor, the process will be slow and may even not iterate towards the correct solution, becoming 'stuck' in a local minimum. Problems are also inherent in the inversion of ill- conditioned matrices (i.e. when the ratio of maximal to minimal eigenvalues is large). Despite these pitfalls it is argued that there is sufficient prior knowledge in the relevant applications to make a 'well informed' initial guess meriting the use of the algorithm. In practice, however, the reconstruction algorithm, in comparison with simple backprojection, is very slow.

4.6   RECONSTRUCTION ALGORITHMS—ULTRASONIC FLOW IMAGING

Ultrasonic flow imaging systems require quite different reconstruction algorithms to the electrical sensing systems described in sections 4.4 and 4.5 because of the form in which the measured data is obtained—essentially a set of time delay measurements giving the distances of object-media interfaces from the receiving sensors.

In practice, as described in Chapter 3.6, ultrasound sensors are mounted on the periphery of a pipe and define the cross-sectional plane to be imaged. Typically a sensor will have a wide sector-angle aperture and be able to transmit a pulse of ultrasound, and then receive echo signals arising from reflections or, less usually, signals transmitted by other sensors (section 4.6.2). Such reflections arise in this application from interfaces in pipe contents, due for example, to gas bubbles in a liquid.

Reconstruction of the flow image may need to be done in real-time so that the reconstruction matches the changing flow patterns passing through the image plane. The processing demand is influenced by the cross-sectional size of the pipe, the flow velocity, and the information that can be extracted from a particular ultrasonic transducer arrangement. The subject is addressed in detail in PhD theses by Gai (1990) and Wiegand (1990). This section is based on their work. The reconstruction of a flow image from ultrasound data can use two processes depending on the form of the ultrasound data, namely: reflection-mode and transmission-mode ultrasound. The former is processed with the ultrasound

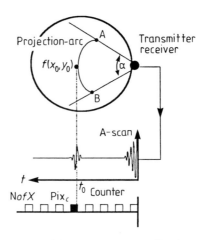

Figure 4.11   Digital data collection.

backprojection method, and the latter is used in a novel way to reduce noise and hence improve the backprojected image (section 4.6.2).

4.6.1   Backprojection with reflection-mode ultrasound

In reflection-mode an ultrasound transducer is firstly used to transmit a short pulse of acoustic energy into the medium, and secondly to receive the resulting reflected signal as a function of time. Represented graphically, the output signal (or *echo*), in reflection mode is called an A-Scan, and provides information relating echo strength against delay time. Knowledge of the ultrasound velocity in the medium allows the time axis to be calibrated as range.

Unlike medical A-Scans, where echo signals often overlap and vary in amplitude due to the complexity of the human body, an A-Scan generated from a 2-component (gas/liquid) flow regime contains mainly discrete echoes. A 2-component flow cross section can be regarded as a binary-field, where there is near zero reflectivity within the liquid phase and high reflectivity at interfaces to the gaseous phase. An echo peak marks the delay time and yields range information, provided there is constant ultrasound velocity in the medium. Figure 4.11 illustrates the relationship between an echo contained in an A-Scan and the resulting digital range information. The reflector at a position $f(x_0, y_0)$ causes an echo with delay time t_0. Now following the treatment by Weigand and Hoyle (1989), the corresponding spatial information d_0, which is the distance to the reflector, is derived from:

$$d_0 = \frac{t_0 c}{2} \tag{4.23}$$

where c is the velocity of sound in the fluid.

For an image array with D_r pixels representing the diameter of the pipe, d_0 can be directly translated to pixel count N_r:

$$N_r = \text{Truncate} \left(\frac{d_0 D_r}{2R} \right) \tag{4.24}$$

where R is the radius of the pipe.

The width of a pixel is calculated as:

$$W_r = \frac{2R}{D_r}. \tag{4.25}$$

The amplitude information of an echo is not used directly in the system described here: it is assumed that any echo above the noise level originates from a liquid/gas interface such as a bubble. A complex echo signal is therefore interpreted as a square pulse with a width corresponding to W_r and unity amplitude.

It is assumed that the reflector is located on an arc determined by the range information, the sector angle aperture α, and the position of the transducer (see figure 4.11). This arc is called the *projection-arc* and in order to reconstruct the image the value unity is backprojected over this arc. Repeating this process for all transducer positions around the pipe produces multiple projection- arcs and their intersections define the reconstructed object(s).

The number of sensor positions around the pipe has a marked effect upon the quality of a reconstructed image. For example Norton and Linzer (1979) have shown that, with a limited number of sensors the filtered backprojection method results in spurious objects being 'detected'. As would be expected these effects progressively degrade the quality as the number of transducer positions decreases. Most of the work reported in the literature has used large numbers of projection measurements, often in excess of 100. Moshfeghi (1986) used up to 360 sensor positions to produce images; Dines and Goss (1987) illuminated the object from between 120 and 360 positions.

A relatively high number of positions is acceptable for non- destructive testing or medical applications, where the data capture time is not critical. In these applications a single probe can be moved around the object in incremental steps. This approach has relatively low cost and avoids problems arising from sensor differences. It is not feasible, however, for flow imaging applications, where real-time image generation is of prime importance. The data acquisition time of an ultrasound flow imaging system is:

$$t_{\text{acq}} = N \left[\frac{4R}{c} + t_{\text{rev}} \right] \tag{4.26}$$

where R is the radius of the pipe, c is the speed of sound in the medium, and N is the number of sensors around the pipe.

A time of t_{rev} must be allowed between successive pulses for reverberations to decay below the noise level, otherwise interference between consecutive pulses corrupts the echo data. For example we assume that reverberations delay in

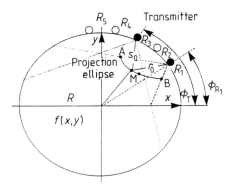

Figure 4.12   Backprojection over a projection-ellipse.

twice the time for a pulse to cross the diameter of the pipe, then t_{rev} can be assumed to be equal to the A-Scan recording time ($t_{scan} = 4R/c$).

For example, to image a water pipe of 137 mm radius with 360 transducer positions results in a data acquisition time of 263 ms according to equation 4.26. Thus the data for 3.8 image frames can be captured in 1 second, which is unacceptable for real-time flow imaging. As a general rule the number of sensor positions needed for a successful reconstruction should be kept as low as possible.

Imaging systems where several sensors receive each transmitted pulse are therefore more appropriate for flow imaging; a scan with multiple receivers has the same data acquisition time as that for a single active receiver, but the amount of data obtained is increased. Echoes received by transducers to either side of the transmitter are the result of contributions from reflectors which lie on an ellipse rather than an arc, where the observed time of flight is constant along the elliptical path.

The ellipse, referred to as the *projection-ellipse*, is defined by its two foci and the major axis. Figure 4.12 illustrates the sensor configuration, where one sensor transmits a pulse and 5 sensors (including the transmitter) receive echo signals concurrently. For example, the locations of receiver R_1 and transmitter R_3 are the foci of a projection-ellipse. The co-ordinates are calculated from the angular positions (ϕ_T and ϕ_{R1}) and the pipe radius R. The major axis is related to the delay time, t_0, for an echo by:

$$\text{major axis} = s_0 + r_0 = t_0 c \qquad (4.27)$$

where s_0 is the distance between transmitter and reflector, r_0 is the distance between reflector and receiver, c is the speed of sound in the medium.

The pixel count Pix_c is determined by equation (4.24), where the length of the major axis is substituted for d_0 in the numerator. For the reconstruction, a projection-ellipse is traced from point A to point B, which are determined by the sensitivity width of the receiver.

An enhancement in the ability of the system to predict the source direction of an echo along the projection arc (for a single active receiver), or ellipse (for multiple active receivers), is achieved when the receiver(s) have *multiple segments*. Such sensors can be constructed where the active sensing surface is split into a small number, say 3, separate segments, each having 1/3 of the angular field of view of a single segment sensor. The increased data from the multiple segments may be employed through modifications to the backprojection algorithm, which in effect add directional information and for example would allow the projection-ellipse from A to B to be split into 3 sectors.

4.6.2 Transmission-mode ultrasound for flow imaging

A further increase in the amount of information available to the reconstruction process, without increasing the number of sensor positions, is made possible by the combination of both reflection and transmission-mode ultrasound data. Due to the assumed binary nature of the medium to be imaged, transmission-mode ultrasound can be integrated into the flow imaging process as follows. The delay times, for direct transmission between each sensor and all other sensors can be calculated for a given sensor arrangement. A received pulse which has been directly transmitted can therefore be distinguished from a reflected pulse, which must have a longer delay time. If a directly transmitted signal is detected it can be concluded that there is no object between the transmitting and receiving sensors. The data acquisition time given by equation (4.26) does not increase, because the receivers for both reflection-mode and transmission-mode are active simultaneously.

For a reconstruction with additional transmission-mode data, one would intuitively set all pixels along the direct paths to zero (i.e. path not blocked). This approach involves two major disadvantages.

Firstly, the reconstruction processing of the reflection-mode data must be completed before the transmission-mode data can be processed or, alternatively, the reflection-mode reconstruction must include a check on each pixel to determine whether it has been set to zero through the use of transmission-mode data. This introduces either the extra overhead of having to buffer the transmission-mode data until all reflection-mode data has been collected and processed, or the inclusion of an extra conditional operation in the processing of each pixel, which is time consuming.

Secondly transmission paths close to a bubble object can intersect pixels which describe the bubble interface, and hence reduce valid reconstructed pixels to zero, as illustrated in figure 4.13.

4.6.3 The reconstruction processes

An incremental line tracing algorithm is used to select the pixels along the lines between the transmitter and the receiver positions (referred to as the *projection-lines*), for transmission-mode data processing. The tracing algorithm employed

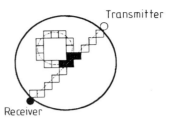

Figure 4.13   Transmission-mode ultrasound close to an object.

was originally developed by Bresenham (1965) for the control of digital plotters. The pixels along the projection-line are selected without the computationally intensive operations of multiplication or division. In order to avoid floating point arithmetic during the execution of the reconstruction algorithm all start and end points for a given sensor geometry are calculated in advance and stored in lookup-tables. This process is referred to as *Line*.

An incremental arc tracing algorithm is used to trace the pixels located on projection-arcs, or projection-arc sectors in cases where multiple segment receivers are employed. The algorithm is again due to Bresenham (1977) being originally developed in this case for the graphical display of circles. The algorithm avoids the need for processing intensive quadratic and trigonometric computations. Multiplication and division operations are also avoided making the algorithm computationally efficient. The process which executes the algorithm is referred to as *Arc*.

A projection-arc is determined by the delay time of an echo, the transducer position, and the sector angle aperture of the receiver. Although the tracing of the pixels lying on projection arcs operates exclusively with integer arithmetic, the calculations of the start, end, and centre points involve time consuming trigonometric operations. The projection-lines are determined in advance and stored in lookup-tables to avoid floating point calculations during the execution of the reconstruction process.

The tracing of projection-ellipses is based on the general equation for conic sections:

$$d(x, y) = \rho y^2 - \sigma x^2 - 2\chi xy - 2uy + 2vx + k. \tag{4.28}$$

The algorithm originates from work by Pitteway (1967, 1985) on the graphical drawing of conic sections with a digital plotter. Once the algorithm is initialized, pixels are traced without multiplications or divisions making the algorithm very efficient. Although the actual tracing of projection-ellipses operates exclusively with integer-arithmetic, the initialisation of the tracing algorithm still requires some floating point operations; once again these are carried out prior to the reconstruction process and stored in lookup-tables. The process which produces the projection-ellipses is referred to as *Elli*.

A flow chart for the image formation procedure is shown in figure 4.14. Two

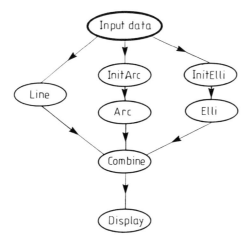

Figure 4.14   Image formation flow chart.

additional processes have been introduced: *InitArc* and *InitElli* which are used for initialization of the lookup-tables for the projection-arcs and the projection-ellipses respectively.

4.7   SUMMARY

The main points of this chapter are:

- Simple back-projection methods of image reconstruction give good results when the sensor field is not seriously distorted by the non-uniform impedance of the object space.

- Iterative reconstruction algorithms may be required to overcome the distortion. These algorithms can be very computer-intensive.

- Flow process images are usually required to yield quantitative information. Care is required to ensure that the quantitative data from the sensors is not corrupted by the image reconstruction process.

- In ultrasonic imaging, the image information obtained by back- projection of reflection mode data can be supplemented by the use of transmission mode data.

5

Image display and interpretation

5.1 INTRODUCTION

Two problems facing the designer of a tomography system, once a reconstructed image has been obtained, are considered in this chapter.
1. Can the quality of the reconstructed image be improved?
2. How can information be extracted from the reconstructed image automatically?

5.2 THE NEED FOR IMAGE PROCESSING

The range of sensing methods suitable for flow imaging was described in Chapter 3. These are intended to provide data from which a cross-sectional image of the flow can be obtained. This type of image is distinctly different from the image obtained by a photographic camera system. Figure 5.1(a) shows a camera imaging system which provides axial information of the flow distribution, but cannot provide cross-sectional information. A typical flow imaging system is illustrated in figure 5.1(b). In this case there is no attempt to obtain axial information but the distribution of the imaging sensors around the pipeline ensures that cross-sectional information can be obtained.

Referring to figure 5.1(b), the analogue signal processing, analogue–digital conversion, and reconstruction algorithms will provide an imperfect image of the flow and it is the purpose of the 'image processing' block described in this chapter to improve the image prior to further analysis or display. Image processing corrects errors in the image by the use of techniques such as the reduction of spurious measurement noise and compensation for the non-uniform sensitivity of the sensing system. This chapter considers the general aspects of image processing; details concerning required accuracy and resolution are discussed in Chapter 6 and the actual methods used to reconstruct the image are described in Chapter 4.

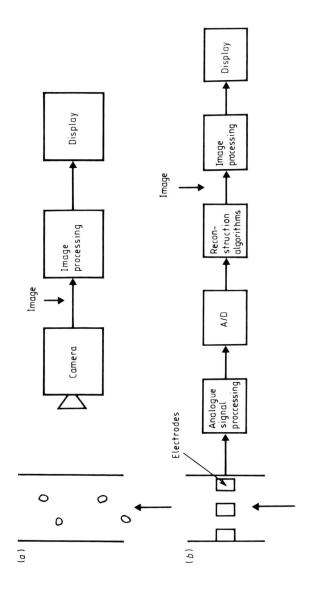

Figure 5.1 Comparison of photography and tomography. (*a*) A camera image provides axial information but cross-sectional information is confused. (*b*) A flow imaging system provides cross-sectional information but no axial information.

5.3   IMAGE PRESENTATION

The flow image will be presented as a monochrome image in terms of a two-dimensional light intensity function $f(x, y)$ (Gonzalez and Wintz 1987) where x and y denote the spatial co-ordinates. The value of f at any point (x, y) is proportional to the brightness (or grey level) of the image at that point. It is sometimes useful to consider a three-dimensional image function with the third axis being the grey level. If we follow the convention of assigning proportionally higher values to brighter areas, the height of components in the plot will be proportional to the corresponding grey level.

The image function $f(x, y)$ is digitized both in spatial co-ordinates and in grey levels. The image is then a matrix whose rows and columns describe the points in the image and the corresponding matrix element defines the grey level at that point. An individual element of this matrix is referred to as a picture element or pixel. The number of pixels is thus the product of the number of rows and the number of columns.

The number of rows and columns (and consequently the number of pixels) changes with the application of the image and the availability of sensing methods to provide the required resolution. There are some computational advantages in selecting square arrays with size and numbers of grey levels that are integer powers of 2, that is

$$\text{The number of pixels } N = 2^n \tag{5.1}$$

$$\text{The number of grey levels } G = 2^m \tag{5.2}$$

So the number of bits required to store an image is given by

$$b = N \times m. \tag{5.3}$$

For example, for a rectangular matrix of 100 pixels (10×10) with binary grey levels we have

$$b = 100 \times 1 = 100$$
$$\text{but for } N = 512 \times 512 = 2^{18} \quad \text{and} \quad m = 8, \ b = 2097\,152 \tag{5.4}$$

The number of pixels N determines the resolution of the image. The value of the parameter m [which is related to the number of grey levels as shown in equation (5.2)] can be related to the amount of phase separation in the flow. For example, with a well separated two-phase flow (e.g. where gas and liquid phases do not mix within an individual pixel zone) a simple binary image as shown in figure 5.2 will be satisfactory so $m = 1$. However, if the two phases may be mixed within an individual pixel zone (corresponding to the case where a significant number of bubbles of the distributed phase are smaller in area than

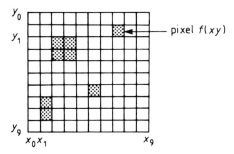

Figure 5.2   Binary image of well separated two-phase flow.

the area of a pixel), then it may be necessary to represent the flow image using intermediate values of grey level.

5.3.1   Pseudo-colour techniques for image display

Colour VDUs can give more readily appreciated presentations of the reconstructed image.   One way of achieving this is to use 'pseudo-colour' techniques to transfer the grey level image (or digital results of reconstruction algorithm) to the pseudo-colour image.   This technique has become very common (Salkeld 1991, Weigand 1990), and a reader can find many up-to-date papers and books on this subject (Gonzalez and Wintz 1987).

One can recognize 2 basic methods (Gonzalez and Wintz 1987):

1. *Density slicing.* This method is similar to a geographic map, using each colour to represent a particular range of values.
2. *Grey level to colour transformation.* This technique is more general and thus capable of achieving a wider range of pseudo-colour enhancement results. Basically, the idea is to perform three independent transformations on the grey levels of any input pixel.   The three results are then fed separately into red, green and blue inputs of a colour monitor.   This produces a composite image whose colour content is modulated by the nature of the transformation function.

5.4   THE GREY LEVEL HISTOGRAM

The grey level histogram is the distribution of the various grey levels as a percentage of the total number of pixels in the picture frame (Pratt 1978). If we assume a digitized picture has a spatial resolution of 128×128 pixels and grey level resolution of 64 levels, the grey level histogram for this particular picture is a method of displaying in what proportion the 64 grey levels are distributed within the picture frame made up of 16 384 pixels. A grey level histogram for a high contrast picture is shown in figure 5.3. This type of histogram is known

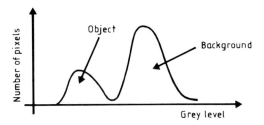

Figure 5.3   Bimodal grey level distribution typical of two-phase flow.

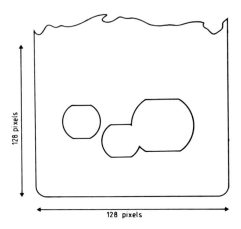

Figure 5.4   Cross-sectional view of physical model of bubble flow.

as a bimodal histogram, because of the two prominent peaks, distinguishing the object from the background.

Figure 5.4 shows a picture of bubble flow (drawn to show boundaries but not showing the contrast range of the actual flow) and figure 5.5 shows the distribution of grey levels (as the percentage) in the whole picture frame (Saeed 1987). The binomial nature of bubble flow can be seen by comparing figure 5.3 with figure 5.5. In the next section (5.5.5) we will examine how filtering techniques affect the grey level histogram.

5.5   ENHANCEMENT OF AN IMAGE USING FILTERING TECHNIQUES

5.5.1   Basic concepts

The main objective of image enhancement is to highlight the flow pattern; that is to improve the separation of objects from the background.   Filtering the

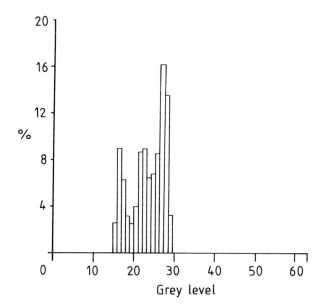

Figure 5.5 Grey level histogram of bubble flow showing bimodal distribution.

picture using a convolution mask results in the modification of the grey level histogram, and by using the correct filtering technique prominent peaks may appear in the grey level histogram display, thus discriminating the object from the background.

A typical 3×3 convolution-filter mask would look as shown below:

$$\frac{1}{N} \begin{bmatrix} w4 & w3 & w2 \\ w5 & w0 & w1 \\ w6 & w7 & w8 \end{bmatrix} \qquad \text{where } N = \sum_{i=0}^{8} w_i. \qquad (5.5)$$

$w0$ to $w8$ can take any real value (though usually an integer). Coefficient $w0$ is centred on the current pixel being examined and this pixel is modified and takes up a new grey value which is dependent on the grey values of the neighbouring pixels and the coefficients $w1$ to $w8$.

Where the sum of the coefficients ($w0$ to $w8$) is zero, then N is replaced by 1, and the convolution mask is called a Laplacian operator.

Consider a 3×3 area of pixels in a matrix frame, where the centre pixel is $P0$. The pixel matrix is shown below:

$$\begin{bmatrix} P4 & P3 & P2 \\ P5 & P0 & P1 \\ P6 & P7 & P8 \end{bmatrix} \tag{5.6}$$

where the value of each element $P0 \dots P8$ denotes the physical property (density, permittivity etc) of the material in the object space. In practice this is measured as a grey level so let us suppose that the corresponding grey level matrix for the above pixel matrix is:

$$\begin{bmatrix} g4 & g3 & g2 \\ g5 & g0 & g1 \\ g6 & g7 & g8 \end{bmatrix} \tag{5.7}$$

where $g0$ to $g8$ correspond to the grey levels of pixels $P0$ to $P8$ respectively.

By convoluting the 3×3 convolution-filter mask with the pixel matrix of grey values $g0$ to $g8$, the grey level of pixel $P0$ (i.e. $g0$) is replaced by:

$$g'0 = \frac{\sum_{n=0}^{8} w_n g_n}{\sum_{n=0}^{8} w_n} \tag{5.8}$$

except when the convolution mask is Laplacian, i.e.

$$\sum_{n=0}^{8} w_n = 0 \quad \text{then} \quad g'0 = \sum_{n=0}^{8} w_n g_n. \tag{5.9}$$

This operation is performed on each and every pixel in the matrix, and as a result the new pixel matrix will have grey levels as shown below:

$$\begin{bmatrix} g'4 & g'3 & g'2 \\ g'5 & g'0 & g'1 \\ g'6 & g'7 & g'8 \end{bmatrix}. \tag{5.10}$$

This operation can be extended to operate on the whole picture frame when required. Filtering techniques for image enhancement will now be considered in more detail.

5.5.2 Low pass filtering

This type of filtering is used for suppressing noise. Noise in general has a higher spatial frequency spectrum than normal image components, and may effectively be smoothed by simple low pass filtering (Pratt 1978).

Typical low pass filter convolution masks, W, are shown below:

(a) simple average

$$W = \frac{1}{9} \begin{bmatrix} 1 & 1 & 1 \\ 1 & 1 & 1 \\ 1 & 1 & 1 \end{bmatrix} \qquad (5.11)$$

(b) centre emphasis

$$W = \frac{1}{12} \begin{bmatrix} 1 & 1 & 1 \\ 1 & 4 & 1 \\ 1 & 1 & 1 \end{bmatrix} \qquad (5.12)$$

(c) centre + 4 neighbour emphasis

$$W = \frac{1}{28} \begin{bmatrix} 1 & 4 & 1 \\ 4 & 8 & 4 \\ 1 & 4 & 1 \end{bmatrix}. \qquad (5.13)$$

The convolution mask, W, contains a scaling factor to give the filter a unity gain. This means that there is no change in the mean brightness of the picture frame,. i.e. grey values of pixels do not overflow.

Larger convolution masks of dimensions 5×5 and 9×9 can be used, but require a long computation time. A typical image consisting of a 128×128 pixel array, convolved with a 9×9 mask, will result in $128 \times 128 \times 9 \times 9$ operation (1.3×10^6 operations).

There is a major problem in using an averaging operation because it tends to smooth out the edges in the image, but edges play an important part when the image is to be segmented, i.e. separation of the object from the background. More sophisticated methods of filtering can be used which divide the image into large cells, determining the mean and variance of the grey values within the cells to detect the presence of edges within the cells. Smoothing of those cells with low variance has the effect of preserving those areas that contain edges.

5.5.3 High pass filtering

This type of filter is used to highlight the edges in the picture. It is achieved in a similar fashion to low pass filtering but negative weighting values are used to highlight edges in an image (Pratt 1978).

The image is convolved with a mask of the following form:

$$W = \begin{bmatrix} -1 & -1 & -1 \\ -1 & 9 & -1 \\ -1 & -1 & -1 \end{bmatrix} \qquad (5.14)$$

$$W = \begin{bmatrix} 0 & -1 & 0 \\ -1 & 5 & -1 \\ 0 & -1 & 0 \end{bmatrix} \qquad (5.15)$$

$$W = \begin{bmatrix} 1 & -2 & 1 \\ -2 & 5 & -2 \\ 1 & -2 & 1 \end{bmatrix} \qquad (5.16)$$

It can be seen that with these high pass masks the sum of the elements is unity and, as in the case of low pass filtering, there is no change in the overall brightness of the image.

5.5.4 Laplacian masks

The Laplacian operator is used to detect edge pixels in a picture frame. Typical Laplacian masks (3×3) take the form shown in equations (5.17)–(5.19).

$$W = \begin{bmatrix} 0 & -1 & 0 \\ -1 & 4 & -1 \\ 0 & -1 & 0 \end{bmatrix} \tag{5.17}$$

$$W = \begin{bmatrix} -1 & -1 & -1 \\ -1 & 8 & -1 \\ -1 & -1 & -1 \end{bmatrix} \tag{5.18}$$

$$W = \begin{bmatrix} 1 & -2 & 1 \\ -2 & 4 & -2 \\ 1 & -2 & 1 \end{bmatrix}. \tag{5.19}$$

In this case, the masks are similar to those used for edge enhancement, but they are zero weighted and convolving these masks with the picture frame gives an output of zero for areas of constant brightness.

5.5.5 Comparison of filtering techniques

Figures 5.6(*a, b*) and *c* show the grey level content of the picture (figure 5.4) after filtering using a low pass filter, a high pass filter and then a Laplacian operator respectively. The filter coefficients are as shown below (Saeed 1987):
  (a) low pass filter

$$W = \frac{1}{9} \begin{bmatrix} 1 & 1 & 1 \\ 1 & 1 & 1 \\ 1 & 1 & 1 \end{bmatrix} \tag{5.20}$$

  (b) high pass filter

$$W = \begin{bmatrix} -1 & -1 & -1 \\ -1 & 9 & -1 \\ -1 & -1 & -1 \end{bmatrix} \tag{5.21}$$

  (c) Laplacian filter

$$W = \begin{bmatrix} -1 & -1 & -1 \\ -1 & 8 & -1 \\ -1 & -1 & -1 \end{bmatrix}. \tag{5.22}$$

Comparing the grey level histogram after using the low pass filter (figure 5.6(*a*)) with the original grey level histogram (figure 5.5), it can be seen that this filter does not have much effect on the grey level content in the picture frame. This means that the original picture was 'clear' (i.e. not noisy)

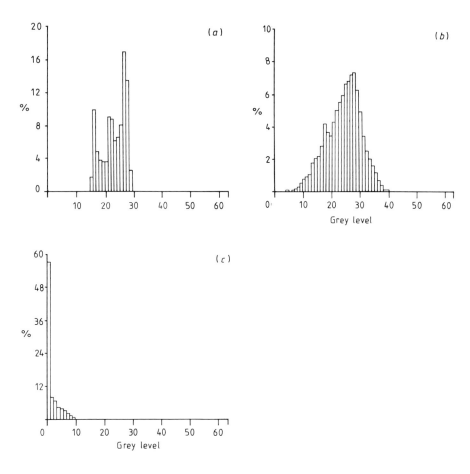

Figure 5.6   Comparison of filtering image of bubble flow (figure 5.4) having unfiltered histogram of figure 5.5; (*a*) low pass, (*b*) high pass and (*c*) Laplacian (from Saeed 1987).

and as such the low pass filter used, whose main purpose is noise removal, had little effect.

When the high pass filter was operated on the picture, it can be seen from the resulting histogram (figure 5.6(*b*)) that the effect of this filter is to stretch the grey level histogram, thus implying a better contrast in the picture, which in effect means highlighting areas where there is an abrupt change in grey level, i.e. edge enhancement.

The effect of the Laplacian operator on the grey level picture is shown in figure 5.6(*c*). Areas with similar grey levels are transformed to low values, pixels with higher grey levels are located at the boundary of the object.

5.6   THE THRESHOLDING TECHNIQUE

5.6.1   Applications to flow imaging

Amplitude thresholding techniques based on the grey level histogram (section 5.4) can be used for separation of objects from the background, thus creating a binary picture from grey picture data. This procedure is appropriate for two-phase flow imaging in cases where the phases are well separated, i.e. where the minimum size zone of any separated phase is larger than the pixel size, so a pixel will be wholly filled by one phase. Pixels falling on phase boundaries will be rounded up or down to the nearest phase level.

5.6.2   Thresholding selection technique

A threshold value is selected from the grey level histogram for the purpose of separating the object from the background. The main problem encountered is to come up with the best thresholding technique so as to obtain an object in the image with the same dimensions as the real object (Weszka 1978).

Most of the thresholding techniques operate on picture frames with well defined object and background. The basis of threshold selection involves choosing a grey level η, such that all grey levels greater than η are mapped into the 'object' label (grey level 1) and all other grey levels are mapped into the 'background' label (grey level 0).

If $g(x, y)$ is the grey level of the pixel point (x, y), and $g(x, y) > \eta$, then the pixel at location (x, y) belongs to the object, otherwise it belongs to the background. An application of this method is described below.

Figure 5.7 shows an example of thresholding using an acoustic echo system to generate flow image data as described in Chapter 3.6. Figure 5.7(a) shows the true image of the test object for reference purposes. Figure 5.7(b) shows the reconstructed image of the same object using four transducer positions and with threshold filtering at 30% of the peak value. This shows a number of spurious peaks and suggests that 4 views are insufficient for efficient flow imaging. Figure 5.7(d) shows the results with twelve transducer positions and threshold filtering at 43% of the peak. In this case, the very simple test object is resolved unambiguously, which suggests that 12 views may be satisfactory. The raw data from the 12-view system before threshold filtering are given in figure 5.7(c). They show the expected low level noise from the reconstruction algorithm. By using a threshold level of 43% of the peak to obtain the output shown in figure 5.7(d), we note that all the noise is less than 22% of the peak value, so that there is a robust margin of safety in picking out the signal corresponding to the test object with this system (Plaskowski *et al* 1987).

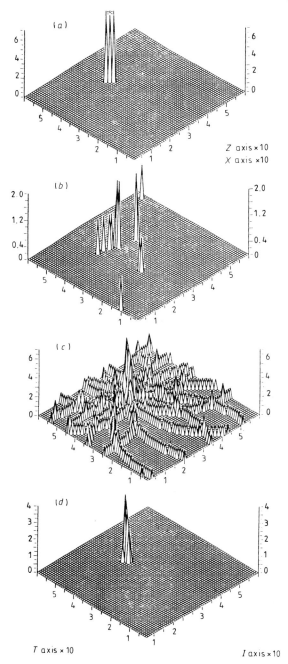

Figure 5.7   Ultrasonic images of 6-mm void in a perspex cylinder. (*a*) Ideal image, (*b*) four transducer views with threshold filtering, (*c*) 12-transducer views without threshold filtering and (*d*) 12-transducer views with threshold filtering.

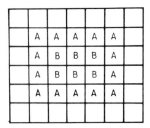

Figure 5.8   Object outlined A is operated on by shrinking algorithm to produce object outlined B.

5.7   SHRINKING AND EXPANSION TECHNIQUES

5.7.1   Introduction

Shrinking and expansion algorithms when used together are very useful routines for the removal of noise, and they also provide a method for separating objects that appear joined together (Pratt 1978). They are also known as erosion and dilation techniques.

5.7.2   Shrinking

Shrinking is an operation which reduces the size of an object whilst preserving the shape (as far as resolution will allow). Figure 5.8 shows how a shrinking algorithm works. The shrinking algorithm uses a defined structuring element, e.g. 3 × 3 pixel region, which is passed over the image. In the case of a binary picture, the pixels in the object are assigned a value of '0', i.e. black pixels, and the background pixels are assigned a value of '1', i.e. white pixels.

The first pass of a shrinking routine will remove all black pixels that come into contact with white pixels, i.e. it will replace these black pixels with white pixels. A pixel in the original object is assigned to the shrunk object if, and only if, the structuring element (centred on the pixel) is wholly contained in the original object. This algorithm can be repeated several times on the picture frame to shrink the object to the required size.

In figure 5.8 pixels with the label 'A' which belong to the original object are shrunk on the first pass of the shrinking routine. The shrunk object is then made up of pixels with label 'B'. Before the first pass of the shrinking program, it can be seen that the pixels with label 'B' are wholly contained in the original object and so are not affected by the first pass of the shrinking algorithm, and thus form part of the shrunk object.

	D	D	D	D	D	D	
	D	C	C	C	C	D	
	D	C	C	C	C	D	
	D	C	C	C	C	D	
	D	D	D	D	D	D	

Figure 5.9   Object outlined C is operated on by expansion algorithm to produce object outlined D.

5.7.3   Expansion

This is similar to shrinking, but here the object is increased in size. The algorithm is similar to the shrinking algorithm, except that an object pixel is in the expanded object if any pixel in the structural element is in the original object. Pixels with label 'C' in figure 5.9 belong to the original object. After the first pass of the expansion routine, pixels with label 'D' now belong to the expanded object.

Shrinking and expansion when used together are very effective algorithms for noise removal. Spurious noise pixels which have an area of one pixel can be eliminated completely from the picture by one pass of the shrinking routine, and then one pass of the expansion routine can be used to rebuild the shrunk object back to its original size. Shrinking and expansion are also used for separating and joining objects respectively.

5.8   MEASUREMENTS ON OBJECTS IN THE IMAGE

5.8.1   Overview

A technique for coding objects, based on their boundaries (the Freeman chain code), is introduced, this reduces the amount of data to be analysed and provides a convenient basis for calculating parameters such as the area, perimeter and centroid of any object detected in the frame.   These parameters are needed for automatic analysis of image data, to provide information required for monitoring and controlling flow processes, for example void fraction and interfacial perimeter can be computed.

5.8.2   Freeman chain code

The Freeman chain coding technique permits the encoding of arbitrary geometric configurations, so as to facilitate their analysis and manipulation by means of a digital computer (Gonzalez and Wintz 1987, Pratt 1978). From the chain code of an object it is possible to calculate the area, perimeter and centre of gravity of the object.

Before an object boundary can be coded using the Freeman chain code, two important steps have to be followed:

(i) the object has to be separated from the background, a procedure known as segmentation;

(ii) the edge structure of an object and hence the boundary has to be recorded in some form. This procedure is known as edge segmentation. There are three general approaches to edge segmentation, namely curve fitting, edge point linking and contour following algorithms (Pratt 1978).

The contour following algorithm will be described in more detail as an example of edge segmentation. This method is mainly applicable to binary images but can be extended to grey scale images. In the case of a binary picture, a 'bug' is considered to start its journey at the bottom left hand corner pixel of the picture frame, and it scans each picture line from left to right until it encounters a black pixel (the initial point of the object). The bug is made to obey a number of rules during its journey from the bottom left hand corner of the picture frame to the top right hand corner. After encountering the first black pixel in an object, the bug obeys the following rules:

- whenever the bug moves into a black pixel, it must turn left for its next move;
- whenever the bug moves into a white pixel, it must turn right for the next move;
- whenever there is a transition from a black pixel to a white pixel, or vice versa, the bug must signal its position during the crossing, i.e. give the co-ordinates of the object boundary.

Using the above procedure the bug zigzags along and across the boundary until it reaches the end of the object, i.e. it returns to the initial point. Figure 5.10 shows an example of the contour following algorithm. Section 5.9 describes how the bug will then proceed to the next object after circumnavigating the previous object.

The Freeman chain code describes the boundary of an object as a series of vectors at different orientations. There are two common ways of defining the Freeman chain code: these are the 4-way coding and the 8-way coding techniques. (The 4-way coding is based on the edge detection procedure described above. The 8-way coding involves an additional step to determine the edge direction.) Figures 5.11(a) and 5.11(b) show the 4-way and the 8-way vector respectively; for example, using these the object in figure 5.12 is coded in a clockwise sense, using both Freeman 4-way (figure 5.12(a)) and 8-way (figure 5.12(b)) coding techniques. There are two distinct advantages of the chain code transformation, which are:

(i) the description of the object is reduced considerably, i.e. in the above

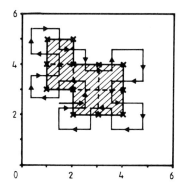

Figure 5.10 Contour following with initial point at (2,2) bug zig-zags along boundary, crossing points on boundary are recorded.

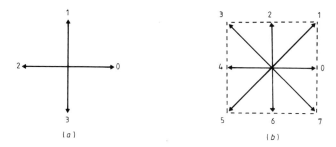

Figure 5.11 Freeman chain code (a) 4-way code vectors (all vectors are of equal length), (b) 8-way code vectors (even vectors are of unit length, odd vectors are of length $\sqrt{2}$).

example the array of 36 pixels is reduced to 10 vectors (8-way coding technique),

(ii) the Freeman data is a one-dimensional string of numbers, thus enabling easy processing of data in comparison with the two-dimensional picture array.

5.8.3   Determining the perimeter from the Freeman chain code

The perimeter of the object can be calculated using either the 4-way or the 8-way chain code. Calculation of the perimeter using the 4-way chain code is very easy, as it is simply the sum of all the codes in the chain code. Here, each code is 1 pixel in length.

Determining the perimeter using an 8-way chain code is similar to the 4-way method, but here the diagonal neighbours are allocated a distance of $\sqrt{2}$ rather than unity, which means that for an 8-way vector, all even vectors (0, 2, 4, 6)

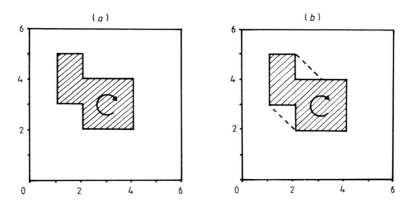

Figure 5.12 Object coded in clockwise sense using 4-way and 8-way coding techniques: (*a*) 4-way chain code is (121103003322); (*b*) 8-way code is (3220706644).

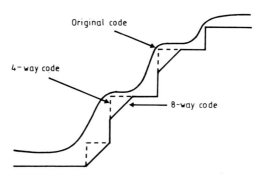

Figure 5.13 Comparison of 4-way and 8-way coding techniques for describing a curve.

are given unit length, with the odd vectors (1, 3, 5, 7) given length of $\sqrt{2}$, i.e.

$$P = N_e + N_0\sqrt{2} \tag{5.23}$$

where P = perimeter, N_e = number of even codes, N_0 = number of odd codes. The perimeter of the 4-way code will be considerably in error for object boundaries which contain non-vertical or non-horizontal edges, since it effectively measures the 'city-block' perimeter of the object (figure 5.13).

Similarly, the 8-way code perimeter will contain some error when compared with the true perimeter. This error will be further compounded if the boundary is corrupted by noise, and in such cases it is worthwhile filtering the boundaries as described in section 5.5.

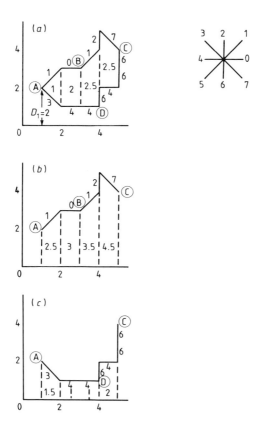

Figure 5.14   Determining area from the Freeman chain code.

5.8.4   Determining area from the Freeman chain code

The area and centre of gravity of an object can be calculated from its Freeman chain code.   The procedure for finding the area is straightforward and is illustrated on figure 5.14 for an object defined by the 8-way vector chain 101 276 646 443.

The complete object is shown in figure 5.14(a). Its area is made up of the area between the curve ABC and the x-axis minus the area between curve CDA and x-axis, figures 5.14(b) and (c) respectively.   The total area is therefore determined by the areas between each Freeman vector and the x-axis. The size of each individual area and whether it should be treated as positive or negative depends on the direction of the vector V and its distance D from the x-axis. The total area A of an object defined by M Freeman vectors is given by

$$A = \sum_{j=1}^{M} \Delta A_j (V_j D_j)$$

Table 5.1   Values of $\Delta A(V, D)$ and $\Delta D(V)$ for all values of V_j

Chain Code V	Area under V_j ΔA	Change in vector distance ΔD_j
0	$+Dj$	0
1	$+Dj + 0.5$	1
2	0	1
3	$-Dj - 0.5$	1
4	$-Dj$	0
5	$-Dj + 0.5$	-1
6	0	-1
7	$+Dj - 0.5$	-1

Table 5.2   Procedure for calculating area of object 101 276 646 443

Chain Code V	Vector distance D	Area under V ΔA	Change in vector distance ΔD
1	2	2.5	1
0	3	3	0
1	3	3.5	1
2	4	0	1
7	5	4.5	-1
6	4	0	-1
6	3	0	-1
4	2	-2	0
6	2	0	-1
4	1	-1	0
4	1	-1	0
3	1	-1.5	1
		Total area $= 8$	

The distance D of each vector depends on the distance D_1 of the initial vector V_1 and the subsequent vector thus

$$D_j = D_1 + \sum_{K=1}^{j-1} \Delta D(V_K)$$

i.e.

$$D_{j+1} = D_j + \Delta D(V_j)$$

where $\Delta A_j(V_j, D_j)$ is the area under vector V_j and $\Delta D(V_j)$ is the change in D due to vector V_j.

The values of $\Delta A(V, D)$ and $\Delta D(V)$ for the 8 possible values of the chain vector are given in table 5.1.

The procedure for calculating the area of the object of figure 5.14(a) is shown in table 5.2.

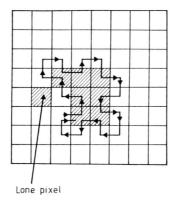

Lone pixel

Figure 5.15 'Lone' pixel not 'visited' by the contour following algorithm.

Note that the area is independent of the initial vector distance D_1 so this parameter is not strictly necessary.

5.9   DETECTION OF ALL OBJECTS WITHIN THE IMAGE FRAME

In order to be able to detect all the objects in the binary picture, a bug starts at the bottom left hand corner pixel of the picture frame and travels along each picture line, pixel by pixel, until it encounters the first black pixel.

The first black pixel forms the starting point for coding of the first object (boundary coding), i.e. the first initial point IP1. The contour following algorithm (section 5.8.2) is then implemented, and the corresponding Freeman code is obtained for the next initial point IP2, i.e. the start of the next object in the picture frame, and then this object is coded. This process is repeated until the 'detector' reaches the end of the frame (Gonzalez and Wintz 1987).

Conventional methods for coding all objects in the picture frame remove from memory the complete object that has already been coded, so that it is not coded again. This is a slow process that could be overcome by intensity coding (i.e. giving it a different numerical value) the boundary pixels of an object that has already been coded. This means that whenever a pixel of intensity coded value is encountered, it is known that this pixel forms part of the boundary of a previously coded object. The bug is programmed to avoid these boundary pixels by jumping over these objects. The task of the bug is to find a black pixel that is in no way connected to an already coded object, and if such a pixel is found this forms the initial point for the coding of a new object.

A situation can arise where a pixel on the boundary of an object is connected in such a way that the contour following algorithm does not 'visit' this pixel during its journey. An example of this is shown in figure 5.15. The 'lone' pixel confuses the bug and leads it to believe that it is an initial point for a new object. This gives a second starting point for the object and results in coding the same

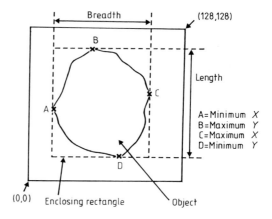

Figure 5.16   Dimensions of enclosing rectangle.

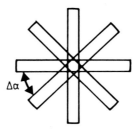

Figure 5.17   Blurring of an object point caused by backprojections.

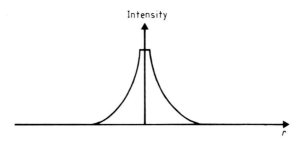

Figure 5.18   Blurring function of an object point.

object, from a different starting point. This shows up in the final result, when more objects than are visible in the picture are detected. As a result of this false starting point, the total area occupied by the objects can be greater than the area of the picture frame. To overcome this false starting point, every initial point (i.e. the pixel value) is compared with the pixel value of its 8 nearest neighbours; if these have not already been coded, then it is an assurance to the bug that the starting point is a genuine initial point for a new object.

Sometimes one can use a simplified method—the enclosing rectangle—to

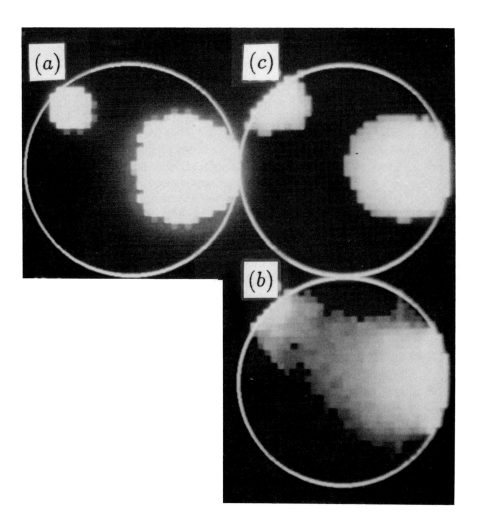

Figure 5.19   Image is improved by correction for sensitivity function: (a) simulation of 'ideal image' for 12-electrode system, (b) image reconstructed without correction and (c) with correction.

describe an object. This is a rectangle which passes through the minimum x, minimum y, maximum x, maximum y coordinate of the object (figure 5.16). The length and breadth of the enclosing rectangle give a good indication of the aspect ratio of the object, e.g. for a square object (or circular object) where the length and breadth of the enclosing rectangle are the same the object aspect ratio is unity.

5.10   BLURRING OF IMAGES CAUSED BY BACKPROJECTION

Reconstruction of an object point through superposition of its projections results in the blurring of the object point. Figure 5.7(c) illustrates these blurring effects on the reconstruction arcs used with ultrasound while figure 5.17 shows the blurring for the case of projection along straight lines. If Δa becomes infinitely small we obtain a continuous intensity distribution, which decreases with increasing distance from the object point and the blurring function is the inverse of the distance r. The intensity distribution is centre symmetrical and thus independent of the angle. It should be noted that this applies only to the ideal case, where the number of projections approaches infinity, i.e. where Δa becomes infinitely small. However, in most applications we assume the blurring function to be $1/r$ in spite of a limited number of projections (figure 5.18). Methods of reducing the blurring by convoluting the image with an inverse of the blurring function have been suggested by various authors. An example of using this procedure for electrical capacitance tomography is shown in figure 5.19. The improved resolution obtained by correcting the image by the inverse of the sensitivity function is clearly shown (Xie *et al* 1991a).

5.11   SUMMARY

The main points of this chapter are:
- Filtering and thresholding techniques can improve the perceived quality of an image and can simplify computer analysis of the image.
- There are a range of techniques available for analysing flow images in order to determine interfacial perimeter, void fraction etc.

6

Applications

6.1 INTRODUCTION

Flow imaging situations of industrial interest are concerned with 3-dimensional fields having non-uniform velocities, temperatures or concentrations. Such fields are found in pipelines where quantitative flow measurement is required and also occur in process reactors, mixers etc. where the flow fields are frequently more complex. For flow measurement where velocity and phase concentration vary unpredictably over the cross section it is essential to be able to measure these parameters simultaneously over a two-dimensional space. Simultaneous measurement is also desirable in process vessels where the dynamic behaviour is rarely isotropic, so measurements in the third dimension may be required.

Instruments based on point-sensing probes have already been developed for making experimental measurements of the three-dimensional velocity, temperature and concentration fields in fluids. Examples of such probes are hot wire and hot film sensors (Doebelin 1983) and laser–Doppler anemometry systems (Durrani and Greated 1977). These point-sensing systems can only be used for measurements in stationary fluids or fluid in steady flow because of the long time needed to scan the whole of a two- or three-dimensional space with sufficient precision. These point sensing methods are therefore not suitable for fast-response flow measurement or for measurements in unsteady state processes.

The value of tomographic measurements in industrial pipelines and processes has been appreciated for several years and the techniques developed for medical tomography have been applied to process situations. Predominant among the early investigations has been the use of ionizing radiation X-ray techniques (Finke *et al* 1982). These have involved the usual difficulties in the safe handling of ionizing radiation and the imaging is slow unless strong X-ray (or radio-isotope) sources are used, which are not encouraged for safety reasons. By contrast, the electrical field methods of tomographic imaging (Chapter 3 and table 3.1) are fast in response, relatively inexpensive and safe. The reader is referred to other texts for details of ionizing radiation methods (Abouelwafa *et al* 1980b), whilst in this chapter we will describe some applications of the more recently developed electrical tomography methods.

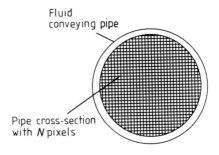

Figure 6.1   The image plane of tomographic flow imaging system.

6.2   ELECTRICAL CAPACITANCE TOMOGRAPHY—PERFORMANCE PARAMETERS

Tomographic imaging using capacitance techniques advanced rapidly in the late 1980s and early 1990s because of energy resource requirements, notably for measurements in the oil industry in Europe and the coal industry in the USA. It is difficult to compare the performance of different systems from case studies and therefore it is useful to prepare some specific definitions of imaging performance appropriate to the electrical tomography systems used for process applications. These performance measures are different to those required for optical lens imaging systems, and have been developed from a study of the 8-electrode and 12-electrode versions of the capacitance tomography system described in Chapter 3, sections 3.4.5, 3.4.6 and 3.4.7; and Chapter 4 section 4.4 (Huang *et al* 1992b). This particular system is designed to measure oil-field flows and was successfully used to image the flow in a 6 inch diameter oil/gas pipeline. However, in view of the random and unpredictable pattern of multicomponent flow, the performance parameters were measured using accurately known physical models of the real flow.

6.2.1   Definitions

The quality of a tomographic flow imaging system can be judged by comparing the reconstructed cross-sectional image (two-dimensional) of a physical model with the actual model. To perform the comparison on the image reconstruction computer, a standard image of the model can be generated, which closely matches the cross-section of the model depending upon the fineness of the image display pixels.

The image plane, representing the cross-section of the flow conveying pipe, is divided into N square image pixels (figure 6.1), and the image of a two-component distribution is defined by assigning an appropriate grey level (or colour code) to each pixel, $G(p)$, for $p = 1, 2, ..., N$. For the standard image:

$$G_s(p) = \begin{cases} G_m & \text{at pixels occupied by the cross-section of the model} \\ 0 & \text{elsewhere} \end{cases}$$

(6.1)

where G_m is a grey level value proportional to the permittivity of the model material. For a two-component flow imaging system, G_m is often chosen to be the maximum grey level of the display (e.g. 255), representing the permittivity of the component with higher dielectric constant.

For the reconstructed image, $G_R(p)$, the grey levels of the N pixels are determined from the set of independent capacitance measurements via the image reconstruction algorithm. An algorithm based on the filtered backprojection method has been developed (Xie *et al* 1992) in which the normalized capacitance measurements, after weighting by their corresponding sensitivity distribution functions, are backprojected onto the pipe cross-section and the resultant image filtered by a thresholding operation.

Let the normalized original reconstructed image be represented by the set of pixels $G_0(p)$ such that $0 \leqslant G_0(p) \leqslant 1$, then the final reconstructed image after threshold filtering is given by the set of pixels $G_R(p)$ where

$$G_R(p) = \begin{cases} 0 & \text{if } G_0(p) < \eta \\ G_m G_0(p) & \text{if } G_0(p) \geqslant \eta \end{cases}$$

(6.2)

and η is the threshold level.

The purpose of threshold filtering is to eliminate the low grey-level noise present in an image. It has been found that the optimum threshold level depends on the component distribution. Numerical investigations have suggested that a suitable dynamic threshold level is given by

$$\eta = (1 - k\mu)\zeta$$

(6.3)

where μ is the average value of the normalized capacitance measurement, ζ is the average value of the normalized original reconstructed image $G_0(p)$, k is a coefficient in the range 0 to 1, typically 0.5. For an ideal tomographic imaging system, the reconstructed image, $G_R(p)$, should be identical to the standard image, i.e

$$G_R(p) - G_s(p) = 0 \qquad (p = 1. 2, ...N).$$

(6.4)

In practice, this rarely happens and the reconstructed image always differs from the standard. The departure from the ideal situation can be characterized by the difference image:

$$D(p) = G_R(p) - G_s(p) \qquad (p = 1, 2, ...N)$$

(6.5)

The difference image is a two-dimensional graphical presentation, and in practice it is convenient to use some simple quantitative values to describe the quality of an imaging system. These criteria are defined in the following subsections.

A. Spatial and permittivity errors. When imaging a physical model in the pipe, the reconstructed cross-section of the model may differ from the standard in area, average grey level, shape and position (e.g. centre of gravity). The differences in area, shape and position can be classified as spatial errors, whereas that in grey level is regarded as permittivity error. Here we define the *spatial image error* (SIE) of the system using the following expression :

$$\text{SIE} = \frac{\sum_{p=1}^{N} |G_\text{B}(p) - G_\text{s}(p)|}{\sum_{p=1}^{N} G_\text{s}(p)} = \frac{\sum_{p=1}^{N} |G_\text{B}(p) - G_\text{s}(p)|}{N_\text{S} G_\text{M}} \tag{6.6}$$

where $G_\text{B}(p)$ is the 'binary' reconstruction image, and is defined as

$$G_\text{B}(p) = \begin{cases} 0 & \text{if } G_0(p) < \eta \\ G_\text{M} & \text{if } G_0(p) \geqslant \eta \end{cases} \tag{6.7}$$

and N_s is the number of pixels occupied by the standard model.

SIE represents the spatial error information seen on the difference image. It contains all the spatial errors such as those in shape, cross-sectional area and position of the reconstructed object. In practice, however, the error in the cross-sectional area of the reconstructed object is often a better representation of the spatial image error, and is easier to estimate than the SIE. The area error (AE) is defined by the following formula:

$$\text{AE} = \frac{\sum_{p=1}^{N} G_\text{B}(p) - \sum_{p=1}^{N} G_\text{s}(p)}{\sum_{p=1}^{N} G_\text{s}(p)} = \frac{N_\text{R} - N_\text{S}}{N_\text{S}} = \frac{N_\text{R}}{N_\text{S}} - 1 \tag{6.8}$$

where N_R is the number of pixels with non-zero grey levels in both reconstructed and binary images.

The *permittivity error* (PE) of the system is defined as the difference between the average grey level, $\overline{G}_\text{R}$, of the reconstructed object and the grey level of the standard model, G_M (see (6.1), divided by G_M, i.e.

$$\text{PE} = \frac{G_\text{R} - G_\text{M}}{G_\text{M}} = \frac{G_\text{R}}{G_\text{M}} - 1 \tag{6.9}$$

where $\overline{G}_\text{R}$ is given by the following formula:

$$\overline{G}_R = \frac{\sum_{p=1}^{N} G_\text{R}(p)}{N_\text{R}}. \tag{6.10}$$

Note that only the N_R non-zero values of $G_R(p)$ are actually summed here.

B. *Component fraction measurement error*. The component fraction (e.g. oil volume fraction in an oil/gas flow) is an important parameter of two-component flow. A major objective of flow imaging is to provide a measurement of this parameter from the cross-sectional image of the two-component flow. In our system, the component fraction, α_R, is calculated by averaging the normalized grey level values of all filtered image pixels over the pipe cross section:

$$\alpha_R = \frac{1}{N} \sum_{p=1}^{N} \frac{G_R(p)}{G_M}$$

$$= \frac{N_R \overline{G}_R}{N G_M} = \left(\frac{N_S}{N}\right)\left(\frac{N_R}{N_S}\right)\left(\frac{G_R}{G_M}\right) \qquad (6.11)$$

$$= \alpha_S(1 + \text{AE})(1 + \text{PE})$$

where α_R is the component fraction of the standard model defined by

$$\alpha_S = \frac{N_S}{N}. \qquad (6.12)$$

The *component fraction measurement error* (CFME) of the system is calculated from

$$\text{CFME} = \alpha_R - \alpha_S = \alpha_S(\text{AE} + \text{PE} + \text{AE.PE}). \qquad (6.13)$$

CFME depends on the combined effect of the *area error* (AE) and the *permittivity error* (PE). However, a small CFME does not necessarily mean that the AE and PE are small because their contributions may tend to cancel each other.

C. *Signal-to-noise ratio*. Due to noise generated by the sensor electronics of capacitance tomography systems (Huang *et al* 1992b), the area and grey level of the reconstructed image and the calculated value of αR fluctuate randomly with time, even when the component distribution in the pipe is stationary. In capacitance tomography, the signal that causes the sensor to respond is the component fraction change. Here we choose the time average value of the component fraction ((6.11), $\overline{\alpha}R$, as the signal and define the signal to noise ratio (SNR) of the system as:

$$\text{SNR} = \frac{\overline{\alpha}_R}{\text{RMS}\,(\alpha_R(t))} \qquad (6.14)$$

where $\text{RMS}(\alpha_R(t))$ is the root-mean-square value of the time dependent variable $\alpha_R(t)$, representing the amplitude of the random fluctuation.

D. *Input signal resolution*. Resolution is a term frequently used for describing the performance of an imaging system. For an image display, it means the spatial resolution determined by the size of image pixels. For a measurement system, on the other hand, it means the smallest input signal that will cause an identifiable output change (Doebelin 1983). With the presence of noise, an output change is identifiable only if it is clearly above the noise level. Therefore the resolution of a measurement system is related to the noise level of that system. For a capacitance tomographic flow imaging system, which can be regarded as a measurement system with the component fraction α_S as its input and reconstructed image as output, we define its *input* to be the reconstructed image. This criterion represents the capability of the system to detect the presence of small changes in component fraction, and it should not be confused with the spatial accuracy of the system which has been defined by (6.6).

In this work, the reconstructed image of a dielectric object (in an empty pipe) is regarded as definite (identifiable) if its *signal-to-noise ratio* (see (6.14) is greater than 3. In determining the *input signal resolution* (ISR), only the simple situation where an object with higher permittivity appears in a homogeneous lower permittivity background (e.g. a plastic test rod in an empty pipe), is considered. In this case, if the reconstructed image has an SNR of 3, then the ISR of the system at that point on the cross section is equal to the area-permittivity product of the test rod (the area of the rod is normalized against that of the full pipe).

Two more definitions can be derived from the above definition . They are:

(1) *Spatial signal resolution* (SSR)—given the differential permittivity (the permittivity difference of the two components) of the test object, the smallest increment in its cross-sectional area (usually from zero) that can be identified (SNR = 3) on the reconstructed image. For two-component flows, the permittivity values of both components are usually known and in the tests reported in section 6.2.2, the permittivity values of the test models are also known. Therefore, the term *input signal resolution* usually means the *spatial signal resolution* and is often expressed as a fraction of the pipe cross-sectional area or as a fraction of the pipe diameter.

(2) *Permittivity resolution*—given the cross-sectional area of the test object, the smallest increment in its differential permittivity (usually from zero) that can be identified on the reconstructed image. This definition is useful when investigating the sensitivity of an imaging system to component permittivity changes.

It should be noted that all the criteria defined above are position dependent. For instance, the SSR at the pipe centre is different from that near the pipe wall. Therefore, it is meaningless to quote a criterion value without mentioning where on the pipe cross-section it applies.

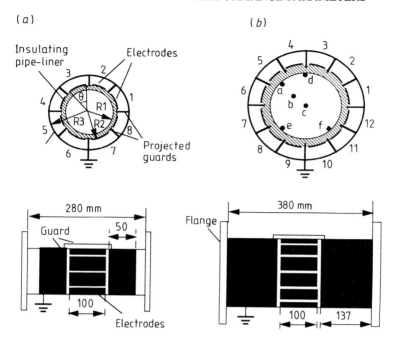

Figure 6.2 Primary sensors for the tests. (a) 8-electrode sensor: $R_1 = 42.5$ m, $R_2 - R_1 = 5$ m, $R_3 - R_2 = 5$ m, $\Theta = 39°$, pipe-liner: Perspex, length of projected guard: 5 mm; (b) 12.-electrode sensor: $R_1 = 77$ m, $R_2 - R_1 = 15$ m, $R_3 - R_2 = 7$ m, $\Theta = 25.6°$, pipe-liner: potting rasin and Perspex, length of projected guard: 9 mm.

6.2.2 Case-study: evaluation of 8- and 12-electrode capacitance tomography systems

The capacitance flow imaging system, described in Chapter 3 (sections 3.4.5, 3.4.6. and 3.4.7) was used for the tests. Two different primary sensors were used for the tests. One had 8 electrodes on a 75 mm Perspex pipe (figure 6.2(a)) and the other had 12 electrodes on a 150 mm Perspex pipe (figure 6.2(b)). The length of the electrodes along the pipe axis was 100 mm in both cases. This length is considered necessary to provide the measuring electrodes with adequate sensitivities (which are proportional to the electrode area). Projected earthed guards are used in both sensors to reduce the standing capacitance between the adjacent electrodes. For the 12-electrode sensor, these guards extend 2 mm into the insulating pipe liner, whereas in the 8-electrode sensor, the guards are flush with the liner surface. The guards at both ends of the electrode section are used to protect the electrodes from the interference of external electrical fields. Their length is no less than the outer diameter of the insulation pipe, R_2 (figure 6.2). The effects of the finite electrode length and the end guards on a two-dimensional imaging system have been investigated and are described by Huang *et al* (1992a).

The sensor electronics has a typical RMS noise level of 0.08 fF, which equals the smallest sensor capacitance change caused by an approximately 0.5% component (permittivity = 3) fraction change (area change) at the pipe centre. The sensor electronics provides a data capture rate of 100 frames (6600 measurements) per second for the 12-electrode system and about 160 frames per second for the 8-electrode system. The image reconstruction algorithm used is based on the filtered backprojection principle. In this the normalized measurement values (normalized between empty-pipe and full-pipe capacitances), weighted by their corresponding sensitivity distribution functions calculated using the finite element method, are backprojected onto the pipe cross-section and the resultant image is filtered by a threshold operation equation (6.3). For the results quoted below, the coefficient k in (6.3) is chosen as 0.5. The reconstructed image is displayed on an image plane (figure 6.1) consisting of 3228 pixels ($N = 3228$). (The image is actually displayed using 64×64 pixels, with 868 pixels outside the pipe).

The physical models simulating different flow patterns can be divided into three types. The first of these are cylindrical rods of plastic (permittivity = 3), which when placed at the pipe centre simulate core flow patterns of different volume fractions. The second are plastic tubes with different wall thickness simulating annular flow. Stratified flow with different volume fractions is simulated by appropriate levels of oil (kerosene, permittivity = 3) in a horizontally laid sensor pipe section.

Extensive tests on the primary sensors shown in figure 6.2 are reported in a paper by Huang *et al* (1992b). The principal conclusions drawn are:

(1) For an object with a permittivity of about 3 in an empty pipe, the minimum size that the imaging system can resolve is 0.2% of the pipe cross-section near the pipe wall and 2.0% at the pipe centre. These are the best *input signal resolution* values that can be achieved with the current signal to noise ratio. For more complex distributions, the resolution might not be as good.

(2) The system errors such as the spatial image error and the permittivity error are strongly dependent on the position and flow pattern. Tests with a 3% rod, near the pipe wall showed that the signal to noise ratio, area error and permittivity error were improved by factors of 6, 3 and 1.33, respectively over the corresponding values at the pipe centre. (Typical results are shown in figure 6.3).

(3) Discrete objects formed by the low permittivity component tend to be obscured by the high permittivity component surrounding them. Similarly objects near the pipe centre tend to be masked by any high permittivity component present near the pipe wall. This is the 'dielectric screen effect'. Due to this effect, objects of lower permittivity in a high permittivity component are more difficult to detect than vice versa. The system errors are particularly large when gas bubbles at the pipe centre are imaged (the minimum reconstructable bubble size is around 20%). The fact that the

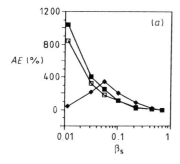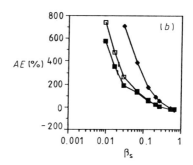

Figure 6.3 The cross-sectional area error (AE) of the test rods at three positions in the pipe (a (–□–), c (–◆–), d (–■–); see figure 6.2(b)) versus the rod area fraction β_s. (a) For 8-electrode configuration. (a) For 12-electrode configuration. (Note in the legend on the right, AE(a) is for the AE at position a, and so on).

bubbles can be detected by the measurement circuit suggests that this effect may be reduced by using image reconstruction algorithms which are more accurate than the linear backprojection method.

(4) When component fraction is estimated from the reconstructed image, the errors can be large for all types of simulated flow regime (all at about 17% maximum). When component fraction is estimated from the average of the 54 normalized capacitance measurements (excluding those from the neighbouring electrode pairs), the errors are significantly reduced (for stratified and annular flows to 5% maximum, for core flow to 12% maximum). In order to improve the accuracy further, either more accurate image reconstruction algorithms or weighted average of the capacitance measurements will have to be used.

(5) Increasing the number of electrodes reduces system errors such as the spatial imaging error, particularly in the areas near the pipe wall. However, the SNR and ISR deteriorate as the number of electrodes increases. When for example the number of electrodes is increased from 8 to 12, the AE for a 3% rod in the near wall area is reduced by about 30%, and the PE by about 15%. However at the pipe centre, the situation is little improved and may even be made worse due to the reduced signal to noise ratio (about half of the original value). This agrees with the theoretical predictions by Seagar *et al* (1987). The use of 12 electrodes instead of 8 may be justified by the reduction of the errors in the near wall area and by the possibility of further reducing the noise level of the sensor electronics.

(6) The images obtained with this system represent the component distribution averaged over the finite length of the electrodes. Reducing this length results mainly in a reduction in the signal-to-noise ratio of the system.

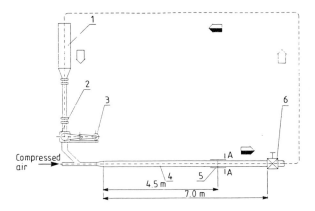

Figure 6.4 Pneumatic conveying system: 1, storage container; 2, rotary feeder; 3, DC motor; 4, test section; 5, capacitance section; 6, speed control valve.

6.3 ELECTRICAL CAPACITANCE TOMOGRAPHY FOR IMAGING GAS/SOLIDS FLOWS IN PNEUMATIC CONVEYERS

Monitoring of two-component flow structures can be helpful for the design and operation of pneumatic transport systems. Examples include the design of transportation systems for fragile foodstuffs, the optimization of solids distribution in pipelines up-stream of flow splitters, and the operation of solids transport pipelines at optimum velocity to minimize energy consumption.

The behaviour of particles in pneumatic conveyors has been widely investigated using various visualization techniques and there is a substantial body of knowledge on critical conveying velocities and on the flow transition boundaries in solids transport (Chapter 2). In practice, it is difficult to operate systems in a satisfactory way using predetermined design data, because the fluid dynamic behaviour of the solid particles may change substantially with slight variations in particle shape, size and distribution. Therefore it is useful to consider on-line techniques which can directly measure the flow pattern in a pneumatic conveyor, on which the effective operation of conveying is highly dependent.

6.3.1 Experimental arrangement

The 8-electrode electrical capacitance tomography system described in Chapter 3 (sections 3.4.5, 3.4.6 and 3.4.7) was installed on the experimental plant shown in figure 6.4. The plant includes a horizontal glass pipe, internal diameter 80 mm and external diameter 96 mm. The system is operated at sub-atmospheric pressure and the solids mass flow rates adjusted by a variable speed turntable feeder. The solids being conveyed during these experiments was rape seed of almost spherical shape with 1.8 mm mean radius and 1210 kg m^{-3} density.

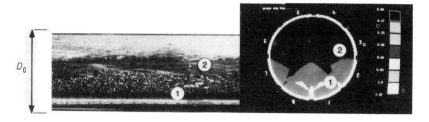

Figure 6.5 Photographs of pipe and image showing (1) dense bed at bottom of pipe and (2) rarified gas/solids mixture in upper part.

A wide range of flow patterns can be generated by varying the feed rate and gas velocity. The pressure drop is measured at 3 m intervals along the conveying section using U-tube manometers. The capacitance electrodes are 190 mm long and are installed at a distance of 4.5 m from the solids feeder, to allow the flow pattern to become essentially fully developed.

6.3.2   Results and discussion

A typical tomographic image of rape seed being conveyed is shown in figure 6.5. Since the capacitance measurements are related to the mean permittivity of the material between the imaging electrodes, it is possible to calibrate the grey level of the image in terms of the dielectric content of the pipe cross-section. The permittivity of the rape seed is known ($\varepsilon_r \approx 3$) for the particular seed and moisture content) and hence the porosity can be estimated from the equation

$$\text{porosity} = \frac{\text{volume of air in the pipe}}{\text{volume solids+volume air}}.$$

The porosity for various grey levels of the dielectric image is shown in figure 6.6, which also indicates the ranges of porosity considered suitable for certain conveying situations.

The results of a series of experiments relating solids mass flow rate to the superficial gas velocity for various solids flow rates are given in figure 6.7. Examples of the observed tomographic flow image are shown covering high velocities where flow turbulence causes breakage of the particulates and a high pumping energy cost is incurred, through to low velocities (dense pneumatic transport or dune flow) where, although particle breakage is minimized and energy consumption low, there would be a risk of complete blockage of the conveyer. Previous work (Chang *et al* 1986) has shown that the conveyor should be operated at an intermediate velocity with plug flow which gives relatively little particle breakage and reliable conveying without risk of blockage. Plug flow can be directly observed as an appropriately time changing pattern on the flow imaging display, as shown in figure 6.8.

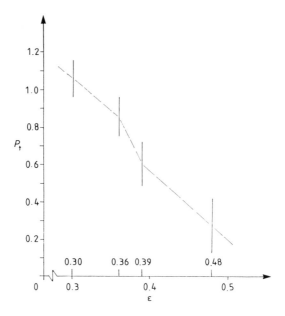

Figure 6.6   Porosity versus grey levels.

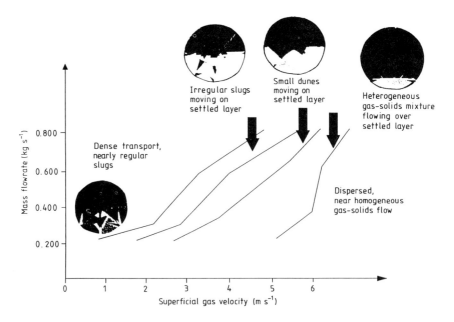

Figure 6.7   Showing superficial velocity versus mass flowrate of air for various solids flows in conveyor.

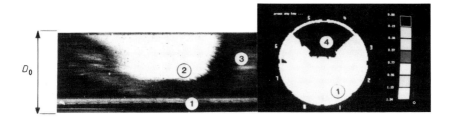

Figure 6.8   Showing image of plug flow seen as that in cross section A–A.

The practical application of electrical capacitance tomography would assist in the following conveying situations:

(i)   Fixed solid mass flow rate requirement. The flow velocity would be reduced to the point where plug flow is observed on the image display, corresponding to minimum energy consumption and minimum breakage for the particular flow rate.

(ii) Transporting a given quantity of solids at minimum energy cost. In this case the conveying compressor would be operated at maximum speed and the solids mass flow rate increased until slug flow was obtained. Thus the compressor would be operating in an efficient mode, and at the same time, the solids mass flow rate would be the highest attainable.

(iii) Accurate splitting of solids flow in pipelines using flow splitting devices (rifflers). A flow image would enable the flow splitting device to be positioned in order to provide the required flows.

(iv) For solids mass flow measurement. The cross sectional density image could be combined with a cross-sectional velocity image in order to accurately measure mass flow in pipelines where the velocity distribution varies over the cross-section, due to saltation and upstream disturbances.

6.4   ELECTRICAL CAPACITANCE TOMOGRAPHY FOR IMAGING GAS FLOW IN FLUIDIZED BEDS

Fluidized beds involve the passage of gas through a layer of solid material for the purposes of heat or mass transfer. Applications include fossil energy technologies involving fluidized bed combustion, coal gasification and chemical reactors. The fluidization behaviour can be characterized by the way in which the gas passes through the coarse particle system of the fluidized bed, a subject which has been extensively investigated at the Morgantown Energy Technology Center of the US Department of Energy (Halow and Fasching 1988, Fasching and Smith 1990).

The Morgantown work includes investigations of 'cold operating' fluidized beds having nylon and plastic spheres to simulate the fossil fuel materials that would be encountered in practice. A diagram of the experimental system for

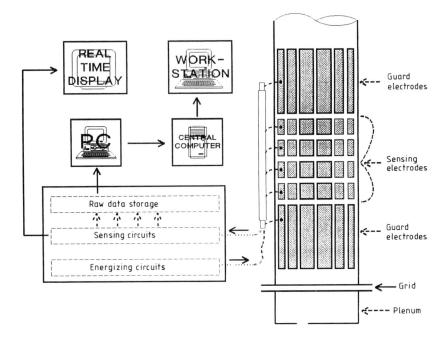

Figure 6.9 The Morgantown energy technologycentre system for imaging gas flow in fluidized beds (Halow and Fashing 1988).

capacitance imaging is shown in figure 6.9. Four rings each having 16 individual sensing electrodes are mounted flush with the walls of the fluidized bed. These electrodes are energized in pairs to generate views based on the distribution of dielectric constant in the bed, which are then reconstructed by solving multi-dimensional simultaneous equations.

The raw images are then transformed onto the pixel definition map similar to that shown in figure 6.1. This polar co-ordinate grid provides the basis for extensive analysis of the images in terms of the size of gas voids in the bed and the velocity of movement.

6.5   FLOW REGIME IDENTIFICATION USING ELECTRICAL CAPACITANCE TOMOGRAPHY SENSING SYSTEMS

Two-component flow can take place in a variety of different flow regimes (figures 2.3 and 2.4) and knowledge of the flow regime can be of great value in industrial processes. For example in oil well risers, changes from bubble to slug flow can suggest the possibility of extremely long high pressure slugs occurring that might overload separation equipment. In heat exchangers the onset of core flow (gas instead of liquid contact) could identify a deterioration in the heat exchange performance and possible damage to the heat exchange elements.

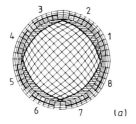 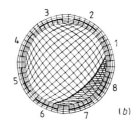

Figure 6.10 Typical finite element grids for preparing reference fingerprint vector. (a) Annular flow, (b) stratified flow.

In principle, flow regime identification should be possible by using conventional image analysis techniques. However, Xie *et al* (1989a,b) have suggested a less hardware intensive system that involves taking the raw capacitance data from the sensing system shown in figure 3.4 and identifying the flow regime without the need to reconstruct the image.

The essential feature of Xie's method is to compare the measured capacitance views, numbering 28 in the case of an 8-electrode system (Chapter 3.4) with a set of stored 'fingerprint' vectors representing the various kinds of flow patterns, and hence identify the actual flow regime.

The classification method works as follows. A library of flow regimes is prepared, corresponding to the different patterns shown in figures 2.3 and 2.4 for a range of flow concentrations (e.g. the ratio of the gas flow to the oil flow in a pipeline). Figure 6.10 shows typical examples of flow regimes together with the finite element reference grid which is used for modelling the fingerprint vector. The next stage is to obtain the fingerprint vector.

The measured data for the 8-electrode capacitance transducer system can be described as a vector C of 28 elements c_r where $r = 1, 2, ..., 28$, i.e.

$$C = (c_1, c_2, c_3, ..., c_{28}) \qquad (6.15)$$

the values of c_1, c_2 etc follow the consistent order:

$$
\begin{array}{llllll}
c_1 = C_{12} & c_2 = C_{13} & c_3 = C_{14} & c_4 = C_{15} & c_5 = C_{16} \\
c_6 = C_{17} & c_7 = C_{18} & c_8 = C_{23} & c_9 = C_{24} & c_{10} = C_{25} \\
c_{11} = C_{26} & c_{12} = C_{27} & c_{13} = C_{28} & c_{14} = C_{34} & c_{15} = C_{35} \\
c_{16} = C_{36} & c_{17} = C_{37} & c_{18} = C_{38} & c_{19} = C_{45} & c_{20} = C_{46} \\
c_{21} = C_{47} & c_{22} = C_{48} & c_{23} = C_{56} & c_{24} = C_{57} & c_{25} = C_{58} \\
c_{26} = C_{67} & c_{27} = C_{68} & c_{28} = C_{78} &
\end{array}
\qquad (6.16)
$$

where, for example, C_{12} is the measured capacitance between electrodes 1 and 2.

Typical fingerprint vectors are given in figure 6.11 for the flows shown in figure 6.10. The repetitive form of the fingerprint vector for the annular flow

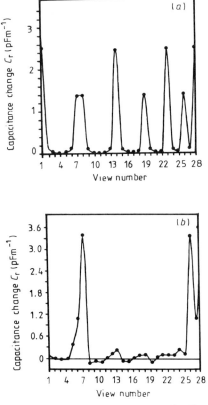

Figure 6.11 Typical reference fingerprints. (*a*) Annular flow, (*b*) stratified flow.

can be clearly distinguished from the non-repetitive fingerprint vector for the stratified flow.

Referring to the flow chart shown in figure 6.12, the classification procedure consists of obtaining the measured fingerprint vector C_m for the actual flow in the pipe, normalizing it and comparing this with the library of normalized reference fingerprint vectors C_r. A difference function λmr is calculated from

$$\lambda_{mr} = |C_m - C_r| / |C_r|.$$

When the values of the different function λ_{mr} has been calculated for the whole library of reference fingerprints, the fingerprint vector corresponding to the lowest value of this difference function will correspond to the actual flow regime.

In his paper Xie points out (Xie *et al* 1989b) that the method described above takes account only of the two-dimensional nature of the flow fingerprint. A more comprehensive classification scheme would also allow for the time history of the fingerprint and would thus include the specific shape function of the flow, for example, slug and elongated churn flow.

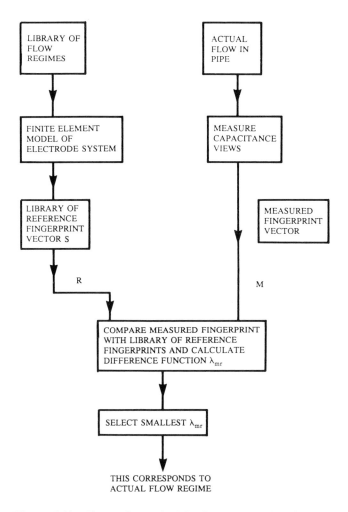

Figure 6.12   Fingerprint method for flow regime classification.

6.6   ELECTRICAL RESISTANCE TOMOGRAPHY FOR MEASURING TRANSIENT CONCENTRATION PROFILES

Electrical resistance tomography (Chapter 3.5) is suitable for measuring component concentration profiles and phase boundaries within pipelines and process vessels where the fluid is electrically conducting. The application described by Abdullah *et al* (1992) used 16 point electrodes around the periphery of a process vessel as shown in figure 3.29.

The materials used for the electrodes depends largely on the process application. Typically the electrode material could be brass, stainless steel or silver palladium alloy made into a bolt or screw-form as shown in figure 6.13.

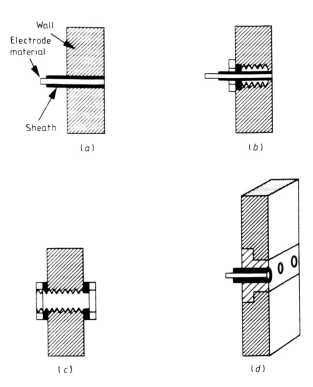

Figure 6.13   Electrode fabrication for EIT system: (*a*) low-pressure vessel, permanently sealed; (*b*) high-pressure vessel, removable electrode (screw-in); (*c*) high-pressure vessel, replaceable electrode (bolt); and (*d*) demountable electrode ring integrated into vessel construction.

The exact method of fabricating electrodes will depend on:
(i)   the corrosive nature of the mixture under investigation,
(ii) abrasion/erosion effects, for example, wear-induced by particulate systems impinging and scouring the vessel walls,
(iii) the process operation environment i.e. temperature, pressure, electrical fire hazards, vessel wall thickness and material,
(iv) the accessibility in terms of spatial and geometric constraints on the vessel and adjacent plant, and the ease of maintaining or, if necessary, replacing electrodes.

Figure 6.14 shows the use of reference imaging technique to image the concentration of sand in water in a stirred vessel. The reference image is reconstructed when the impeller is stirring water alone in the vessel. This image is subtracted from the image obtained when there is sand in the vessel, hence the resultant image shows the distribution of the sand in the water, with the image data of the stirrer suppressed.

Plane 2 @ 280 rpm
(Ref = impeller)

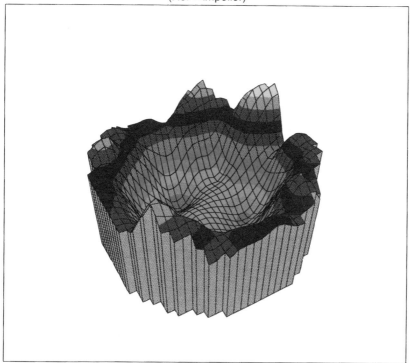

Figure 6.14   Results of 16-electrode EIT system for measuring concentration profile of sand slurry in a 300 mm diameter vessel.   Mean solid diameter 150–210 μm, average weight concentration 15%, reconstruction by quatitative algorithm (Abdullah *et al* 1992), spatial resolution 13%.

6.7   SUMMARY

The main points of this chapter are:
- The potential of tomographic measurements in industrial pipelines and processes is increasing as more successful applications are demonstrated.
- Recently developed electrical and ultrasonic methods of imaging are now being used instead of traditional ionizing radiation methods.
- The ways in which the quality of a reconstructed image can be determined is an area which is not yet fully developed.

7

The future

7.1 THE NEED FOR ACTION

Industrial flows can be imaged by a variety of techniques (table 3.1). Electrical tomography has emerged in the 1990s as an imaging technique capable of meeting many significant requirements—safety, cost and speed. The case studies in this book have been largely concerned with the development and application of electrical tomography.

Perhaps the most significant development in the 1990s has been the realization that only by a pooling of knowledge on a continental or world basis, can the various methods of tomographic technology be properly compared and related to the cardinal needs, which must always be to operate industrial processes at higher efficiency, produce less pollution and be designed to be 'right first time' without excessive capital cost.

This quest resulted in the formation in 1992 of the European Concerted Action on Process Tomography. By 1994 the concerted action is showing how tomographic imaging can develop on a wide front with the prospect of substantial benefits to industrial process operation. Just when this book is being published the Engineering Foundation is launching a worldwide conference on 'frontiers in industrial process tomography' to address the current status and critical future developments needed to apply tomographic technology to process engineering. The stated aim is 'to exploit the recent rapid development in a variety of sensor methods and data processing techniques to provide robust instrumentation to facilitate better control of manufactured products, development and validation of fluid flow and process models, multiphase flow monitoring, and innovative environmental control'.

The basis is established. Let us analyse recent trends and try to 'gaze into the crystal ball'.

Looking at tomography on the widest front, tomographic techniques have been used for some time, although applications to process equipment have only become commonplace recently. Figure 7.1 charts the results of an extensive database search (McKee 1992) showing the number of publications per head from 1970 to 1990 relating to various tomographic methods. As expected,

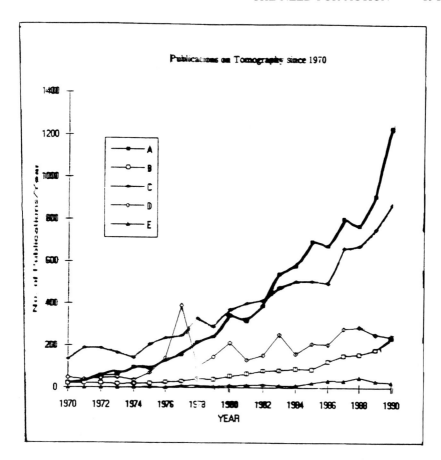

Figure 7.1 Publications on process tomography 1970–1990. A: Radiation based tomography (NMR, x-ray, gamma-ray and PET); B: Impedance (capacitance and resistance) tomography; C: Optical tomography; D: Ultrasonic tomography; E: No of publications related to the chemical/process industry.

radiation methods pertaining to clinical diagnosis and non-destructive testing dominate the literature. Electrical methods also featured quite prominently, in association with their use in geophysical and marine exploration. However, it will be seen that the number of publications in which tomography has been applied to chemical and process industry applications remained small until 1990. Undoubtedly this number is increasing rapidly. The European Concerted Action in Process Tomography (ECAPT) has added a publication rate of 60 per year in its own conference meetings (table 7.1). A survey of the 42 member organizations showed that in addition to the ECAPT meeting publications, they were publishing papers on process tomography in a variety of journals at a rate of about 150 papers per year.

7.2   THE NEW TRENDS

Table 7.1 shows how many organizations, many people and their concerted action is producing a large output of work on process tomography.   The proceedings of ECAPT reflect this (Beck *et al* 1993, 1994a, b). It is impossible here to review the total development of tomography represented at ECAPT. We will however look at a range of techniques, which are relevant to the work of the authors' laboratories.

7.2.1   Electrical capacitance tomography (ECT)

ECT is the most mature of the electrical sensing methods with proven and developing industrial applications in:
- Oilfield technology: oil/gas/water 3-component mass flowmetering (component fraction and velocity distribution imaging). Separator imaging and level measurements.
- Pneumatic transport: gas/solids flow imaging and mass flow measurement.
- Fluidized bed: gas/solids imaging.
- Engine flame-front imaging: High-speed and high contrast-ratio imaging. Basic research is focusing on:
- Capacitance sensing electronic circuitry with even greater sensitivity and higher resolution and SNR for use with primary sensors with more or shorter electrodes. For some applications the circuitry should also be able to measure the conductance component for deconvolution/correction purposes; measurement at a range of excitation frequencies may also be needed.
- 2D/3D mathematical modelling tools for primary sensor design, sensor performance prediction and algorithm development and/or implementation.
- Image reconstruction algorithms taking into account the field distortion caused by high loss or high permittivity components (e.g. saline water).
- Real-time implementation of image reconstruction algorithms (parallel algorithms).

7.2.2   Electrical resistance tomography (ERT)

ERT (referred to as impedance tomography in Chapter 3) is rapidly developing due to numerous process applications associated with conducting fluids. Present work is based on flexible instrument that can be used in many different applications for obtaining both qualitative and quantitative maps of electrical conductivity. Applications include:
- stirred tank reactors,
- centrifugal separators,
- conveying systems,
- voidage mapping in porous beds, sediments and rock core samples,
- tracer migration in porous beds and soils.

Table 7.1 The European Concerted Action in Process Tomography (ECAPT).

	Contributed papers and keynote reviews to ECAPT'94	Other publications on PT August 92 to November 93	Total number of conferences (excluding ECAPT) where PT was presented	Total number of industrial tests on PT performed by academic groups	Total number of tests on PT performed by industrial groups	Number of academic groups	Number of industrial groups	Total number of countries represented
UK	33	48	35	14	52	16	6	
Germany	6	18	19	0	—	4	0	
France	7	38	17	3	—	2	2	
Norway	5	6	8	1	1	1	2	
Others	7	44	9	0	2	6	3	
Total (1994)	58†	154	88	18	55	29	13	10
Previous total (1993)	65	‡	‡	‡	‡	24	13	10
Previous total (1992)	40	‡	‡	‡	‡	15	5	8

† Full referring introduced in 1994.
‡ Data not requested.

Current research involves imaging across dimensions of 1 mm up to 1 m and above using various novel electrode geometries with the ultimate goal of acquiring information in three dimensions. Basic research is focusing on:
- Establishing limits to sensitivity and resolution,
- Use on metal walled vessels,
- Development of 3D image reconstruction algorithms and dedicated computer architectures.

7.2.3   Electromagnetic inductance tomography EMT

The least developed and newest technique. A pre-prototype device has been constructed. We foresee important applications in:
- Metallurgical operations (dense medium separators, scrap sorting etc).
- As an electrodeless replacement for EIT in some process applications and also in imaging the human body (where EMT's penetration of insulating bony structure and voids may be a decided advantage over EIT).
  Basic research is focusing on:
- Sensor design. Rotating parallel field or focused coil or coils. Measurement properties, permeability, eddy current loss or both? Detection by field distortion, Q, or change in resonant frequency? Use of orthogonality of permeability and eddy current data.
- Field modelling.
- Use of in-phase and quadrature signal components for dual modality imaging both resistive and permeability components.
- Use of EMT at high frequencies for imaging water content.

7.2.4   Microwave tomography

Tomographic imaging by microwaves can make use of the diffraction properties of electromagnetic fields, this contrasts with capacitive, resistive and inductive methods of tomography which are based on line-integral effects of their fields. Therefore microwaves will have advantages in imaging objects where the material boundaries, or grain size have dimensions appropriate to the microwave wavelengths employed. We foresee applications in
- Locating pre-emergent knots in timber in order to optimize timber-cutting in sawmills.
- Imaging voids in densely packed dielectric beds (for example for investigating the air distribution which is important for the efficient operation of fluidized bed reactors and heat exchangers). This is usually done on physical simulations for model verification purposes.
  Basic research will need to be focused on
- Optimum choice of wavelength to image specific features in the object space.
- Designing cost-effective multiple view microwave delivery and receiver systems for projection imaging.

- Image reconstruction techniques to effectively combine diffraction, scattering and absorption data.

7.2.5 Multimodality and impedance (dielectric) spectroscopy imaging systems

Multimodality tomographic systems are those in which two or more different sensing modalities are used to locate or measure different constituents in the object space. Such systems are required to provide 'component specificity', for example the use of combined electrical capacitance and ultrasonic tomography systems for oil/gas/water imaging. The most significant problem in such multimodality imaging systems is that of image registration because the field equipotential map can be influenced by the distribution of material in the object space.

A particular sub-class of multimodality systems are the 'inherently multimodality imaging systems'. In these systems only one sensing modality is used, so that the problems of image registration are avoided. The most promising electrical method for securing inherent multimodality is the use of impedance spectroscopy, frequently referred to as dielectric spectroscopy when operating at high frequencies. A project has been carried out to investigate the potential of impedance spectroscopy for differentiating between different materials in liquid/solid mixtures, and is starting to show promising results. We see applications of multimodality and inherently multimodality systems in

- Oil/gas/water imaging in oilfield pipelines and separators.
- High contrast imaging of specific components by frequency-difference imaging.
- Non-destructive evaluation of composite (multilayer) materials.
  Further research needs to be focused on
- Incorporation of impedance spectroscopy in low frequency ERT, high frequency capacitive and microwave, and electromagnetic field tomographic systems.
- Optimum methods of view-scanning to reduce motion-induced errors when using impedance spectroscopy.

7.2.6 Silicon technology in tomographic systems

Tomographic imaging systems use a multiplicity of identical sensors for parallel data collection, together with complex digital systems for sensor control. By constructing the systems on application specific integrated circuits (ASIC's) the size and reliability can be improved and the cost reduced. The latter aspect will be most significant in enabling tomographic type instrumentation to be used for a larger market than that of specialized process applications (we are already considering simplified tomographic systems for locating hidden objects in building cavities, integrity of trench infill material, etc). A recently developed silicon-chip (Eurochip SGS-Thomson process) which shows potential for performing very well in electrical capacitance tomography. A later stage may be to exploit the simple yet massive parallelism possible on a silicon chip to produce lower cost image reconstruction systems. We foresee important applications in

- Tomographic type sensing systems for use in multi-interface level measurement, hidden object location, etc.
- Ultra-sensitive capacitance tomography systems with the input FET gates directly exposed to the sensing field.
- Passive electrodynamic (triboelectric) charge sensors for near-boundary imaging.

Basic research will need to be focused on

- Optimum use of silicon design to maximize signal/noise ratio of the sensors.
- Securing high reliability and long-life when sensors are exposed to the environmental and electrostatic hazards in process equipment.
- Exploiting massive parallelism for image reconstruction, and compensating for image distortion caused by field equipotential shift.

7.2.7   Image reconstruction algorithms

Most process tomography work has concentrated on the use of a linearized version of the back-projection algorithms originally developed for X-ray tomography. These have the advantages of low computation cost and immunity to sensor noise. They simply sum the attenuation coefficients of the pixels along a straight line. In electrical tomography the summation is along the current flow lines, which vary in position because of variations in the material in the object space. This variation in the sensing field causes distortion of the image and reduces the image resolution. Theoretical work has been carried out on the development of iterative and other image reconstruction techniques to reduce the effect of field distortion, which has yet to be fully exploited in process tomography. This should be done and we foresee applications in

- Reducing image distortion in high impedance contrast systems (e.g. oil/water mixtures).
- Improved image registration for multimodality imaging.

Further research will be required on:

- The effect of sensor noise on the reconstruction algorithm.
- The optimum use of parallel computer architecture to reduce image reconstruction time.
- Dedicated algorithms suitable for cost effective realization on dedicated silicon.
- Artificial neural network approach for image reconstruction and pattern identification.

7.2.8   Vector velocity imaging

For full elucidation of process characteristics, it is sometimes necessary to measure the velocity and direction of material movement. Examples occur in multi-component flow measurement in pipelines, assessing the performance of stirred vessels, hydrocyclones etc. In a recent project we have explored the use of the cross-correlation of image information to measure the axial velocity profile in a two-component pipeline. More research is needed in this area which will lead to applications in:

• Quantitative component flow measurement in multi-component flow systems.
• Measurements for process design and operation in mixing and separation processes.

Basic research will be focused on:

• Considering the relative merits of cross-correlation of tomographic view data, sometimes followed by reconstruction, compared with cross correlation of reconstructed image pixel data.
• The relative merits of various computational techniques including digital signal processors, general purpose array processors and dedicated silicon hardware.

7.2.9   Tomographic image analysis

For many applications of process tomography, automatic analysis of the image data is required. Examples of this are:

• Process control.
• Detecting malfunctions in processes.
• Multicomponent flow measurement.
• Flow regime identification (with application to evaporation/condensation processes, flow meter, calibration etc).

Image analysis techniques have been extensively developed for other fields (object recognition, target tracking etc). Specific future research for process tomography is likely to include:

• Tracking the position of specific features in the object space, for example locating the position of the air-core in a hydrocyclone for process control purposes.
• Enhancement of the information from particular parts of the object space by optimum location of sensors and design of reconstruction algorithms.

7.2.10   Process control using tomographic image data

The largest benefits of tomography may come from on-line control using internal measurements obtained from parts of a process where 'other instruments cannot reach'. In 1992 a project on control of hydrocyclone separators using tomographic data was started; the outlook is promising. We foresee a major application in:

- Controlling processes in an optimal way much closer to a boundary constraint (e.g. operating 2 phase conveyors and separators at very high solids fluid phase ratios with consequent energy savings).

Further research will be needed on:

- Image analysis to derive the measured variables from qualitative image data (image co- ordinate analysis), quantitative image data (region of interest location and analysis).
- Modified projection system to enhance information from region of interest.
- Robustness of control using tomographic data.

References

Abouelwafa M S A, Kendall E and John M 1980a The use of capacitance sensors for phase percentage determination in multiphase pipelines *IEEE Trans. Instrum. Meas.* **IM-29** 24–7

Abouelwafa M S A and Kendall E I M 1980b The measurement of component ratios in multiphase systems using gamma-ray attenuation *J. Phys. E: Sci. Instrum.* **13**

Annunziato M and Girrardi G 1987 Horizontal two phase flow: a statistical method for flow pattern recognition *Proc. 3rd Int. Conf. On Multi-Phase Flow* (The Hague)

Asher R C 1983 Ultrasonic sensors in the chemical and process industries *J. Phys. E: Sci. Instrum.* **16** 959–63

Auracher H and Daubert J 1985 A capacitance method for void fraction measurements in two phase flow *Proc. 2nd Int. Conf. On Multi-Phase Flow* (London) pp 425–41

Bair M S and Oakley J P 1992 Comparison of excitation methods for electrical capacitance tomography *Tomographic Techniques for Process Design and Operations* ed M S Beck, E Campograde, R C Waterfall and R A Williams (Southampton: Computational Mechanics Publications)

Baker R C and Hemp J 1981 *Slurry Concentration Meters* (Bedford: BHRA)

Balachandran W and Beck M S 1980 Solids concentration measurement of slurries and sludges using ultrasonic sensors with random data analysis *Trans. Inst. Meas. Control* **2** 181–97; 199–206

Barber D C and Brown B H 1984 Applied potential tomography *J. Phys. E: Sci.Instrum.* **17** 723–33

Barnea D 1987 A unified model for predicting flow-pattern transition for whole range of pipe inclination *Int. J. Multiphase Flow* **13** 1–12

Barnea D and Brauner N 1985 Holdup of the liquid slug in two phase intermittent flow *Int. J. Multiphase Flow* **12** 733–44

Barnea D, Shoaham O and Taitel Y 1982a Flow pattern transition for downward inclined two phase flow; horizontal to vertical *Chem. Eng. Science* **37** 735–40

—— 1982b Flow pattern transition for vertical downward two phase flow; horizontal to vertical *Chem. Eng. Sci.* **37** 741–6

Beard K V and Pruppacher H R 1969 A determination of the terminal velocity and drag of small water drops by means of a wind tunnel *J. Atmos. Sci.* **26** 1066–72

Beck M S, Morris M, Waterfall R C, Williams R A and Campograde E 1992 *Tomographic Techniques for Process Design and Operation* (Southampton: Computational Mechanics Publications)

—— 1993 *Process Tomography—A Strategy for Industrial Exploitation* (Manchester: UMIST)

Beck M S, Morris M, Waterfall R C, Williams R A, Campograde E and Hammer E A 1944 *Process Tomography - A Strategy for Industrial Exploitation* (Manchester: UMIST)

Beck M S and Plaskowski A 1987 *Cross-Correlation Flowmeters: Their Design and Application* (Bristol: Adam Hilger)

Becker E, Hiller W J and Kowalewski T A 1991 Experimental and theoretical investigations of large amplitude oscillations of liquid droplets *J. Fluid Mech.* **231** 189–210

Bopp S, Durst F, Teufel M and Weber H 1990 Volumetric flow rate measurements in oscillating pipe flows with a laser–Doppler sensor *Meas. Sci. Technol.* **1** 917–23

Bresenham J 1965 Algorithm for computer control of a digital plotter *IBM Systems J.* **4** 25–30

—— 1977 A linear algorithm for incremental digital display of circular arcs *Commun. ACM* **20** 100–6

Cascetta F, Della Valle S, Guido A R and Vigo P 1989 A Coriolis mass flowmeter based on a new type of elastic suspension *Measurement* **7** 182–91

Chang J S, Tofiluk W, Myint T A, Hayashi N and Brodowicz K 1986 Time averaged particle fraction and flow patterns in gas-powder *Proc. 4th Miami Int. Symp. On Multi-Phase Transport (Miami)*

Chen Q, Hoyle B S and Strangeways H J 1992 Electric field interaction and an enhanced resolution algorithm in capacitance process tomography *Tomographic Techniques for Process Design and Operations* ed M S Beck, E Campograde, R C Waterfall and R A Williams (Southampton: Computational Mechanics Publications)

Clift R, Grace J R and Weber M E 1978 *Bubbles, Drops and Particles* (New York: Academic)

Collins R 1967 The effect of a containing cylindrical boundary on the velocity of a large gas bubble in a liquid *J. Fluid Mech.* **28** 469–80

Considine M C 1985 *Process Instruments and Controls Handbook* (New York: McGraw-Hill)

Crecraft D I 1983 Ultrasonic instrumentation: principles, methods and applications *J. Phys. E: Sci. Instrum.* **16** 181–9

Delhaye J M 1974 Jump conditions and entropy sources in two-phase systems. Local instant formulation *Int. J. Multiphase Flow* **1** 395–409

de Nevers N and Wu J L 1971 Bubble coalescence in viscous fluids *AIChE. J.* **17** 182–6

Dines K A and Gross S A 1987 Computed ultrasonic reflection tomography *IEEE Trans. Ultrasonics, Ferroelectrics and Frequency Control* **UFFC-34** 309–17

Doebelin E O 1983 *Measurement Systems: Application and Design* 3rd edn (London: McGraw-Hill)

Drew D A 1983 *Continuum Modelling of Two-Phase Flows in Theory of Dispersed Multiphase Flow* ed R E Meyer (New York: Academic)

Durani and Greated 1977 *Laser Systems in Flow Measurement* (New York: Plenum Press)

Dykesteen E and Frantzen K H 1990 The CMI multiphase fraction meter *Proc. Int. Conf. On Basic Principles and Industrial Applications of Multiphase Flow* (24–25 April) (London: BHRA)

Ellul I R and Issa R I 1987 The prediction of interspersed two-phase flow through pipe obstruction *Proc. 3rd Int. Conf. On Mutliphase Flow* (The Hague: BHRA Fluid Engineering)

Fasching G E and Smith N S 1990 A capacitive system for three-dimensional imaging of fluidized beds *Rev. Sci. Instrum.* **62** 2243–51

Finke J R, Vince M A and Jettry C L 1982 Measurement of time-averaged density distribution in horizontal multiphase flow using reconstructive tomography *AIAA/ASME 3rd Joint Thermophysics, Fluids, Plasma and Heat Transfer Conf.* (St Louis)

Franca F, Acikgoz M, Lahey Jr R T and Clausse A 1991 The use of fractal techniques for flow regime identification *Int. J. Multiphase Flow* **17** 445–552

Furness R A 1990 Coriolis meters present and future *Control and Instrumentation* **22** 42–3

Gai H 1990 Ultrasonic techniques for flow imaging *PhD Thesis* University of Manchester

Gai H, Beck M S and Flemons R 1989b An integral transducer/pipe structure for flow imaging *Proc. IEEE 3rd Int. Ultrasonic Symp.* (Montreal: IEEE)

Gai H, Li Y C, Plaskowski A and Beck M S 1989a Ultrasonic flow imaging using time-resolved transmission mode tomography *Proc. IEE 3rd Int. Conf. On Image Processing and Its Applications* (Warwick: Warwick University Press)

Galletti P M, De Rossi D E and De Reggi A S 1988 Medical Applications of Piezoelectric Polymers (New York: Gordon and Breach Science Publications)

Genthe W K 1974 The nuclear magnetic resonance flowmeter, process flow measurement experience *Symposium On Flow Measurements and Control in Science and Industry* ed R E Wendt (Research Triangle Park, NC: Instrument Society of America) pp 849–56

Gimson C 1989 Using the capacitance charge transfer principle for water content measurement *Measurement and Control* **22** 79–81

Gonzalez R C and Wintz P 1987 *Digital Image Processing* 2nd edn (Reading, MA: Addison-Wesley)

Grace J R 1973 Shapes and velocities of bubbles rising in infinite liquids *Trans. Inst. Chem. Eng.* **51** 116–20

Grace J R, Wairegi T and Nguyen T H 1976 Shapes and velocities of single drops and bubbles moving freely through immiscible liquids *Trans. Inst. Chem. Eng.* **54** 167–73

Grant I D R 1975 *Advance in thermal and mechanical design of shell and tube heat exchangers. Vol : Flow and Pressure Drop With Single-Phase and Two-Phase Flow On a Shell-Side of Segmentally Baffled Shell and Tube Heat Exchangers* NEL report 590

Green R G 1981 Capacitance flow transducers for multiphase systems *PhD Thesis* University of Bradford

Green R G and Taylor R W 1986 The instrumentation and control of a wet peening process *J. Phys. E: Sci. Instrum.* **19** 110–5

Halow J S and Fasching G E 1988 Preliminary capacitance imaging experiments of a fluidized bed *AIChE. Symp. Ser.* **86** 41–50

Hammer E A 1983 Three-component flow measurement in oil/gas/water mixtures using capacitance transducers *PhD Thesis* University of Manchester

Hammer E A, Tollefsen J and Olsvik K 1989 Capacitance transducers for non-intrusive measurement of water in crude oil *Flow Meas. Instrum.* **1** 51–8

Harris D B, Llewellyn G J and Beck M S 1976 Turbulence noise and cross-correlation techniques applied to measurements in stacks *Chem. Ind. (London)* 634–7

Hayes D G, Gregory I A and Beck M S 1992 Velocity profile measurement in two-phase flows *Proc. Concerted Action On Process Tomography* ed M S Beck, E Campograde, R C Waterfall and R A Williams (Southampton: Computational Mechanics Publications)

Herman G T 1980 *Image Reconstruction from Projections, the Fundamentals of Computerized Tomography* (New York: Academic)

Hetsroni G 1982 *1982 Handbook of Multiphase Systems* (New York: McGraw-Hill)

Hewitt G F 1982 Flow regimes *Handbook of Multiphase Systems* ed G Hetsroni (New York: McGraw-Hill)

Hewitt G F and Hall Taylor A 1970 *Annular Two-Phase Flow* (Oxford: Oxford University Press)

Hewitt G F, Jayanti S and Hope C P 1990 Structure of thin liquid films in gas liquid horizontal flow *Int. J. Multiphase Flow* **16** 951–7

Huang S M 1986 Capacitance transducers for concentration measurement in multicomponent flow processes *PhD Thesis* University of Manchester

Huang S M, Green R G, Plaskowski A and Beck M S 1988 A high frequency stray-immune capacitance transducer based on the charge transfer principle *IEEE Trans. Instrum. Meas.* **37** 368–73

Huang S M, Plaskowski A, Xie C G and Beck M S 1988b Capacitance-based tomographic flow imaging system *Electron. Lett.* **24** 418–9

—— 1989 Tomographic imaging of two-component flow using capacitance sensors *J. Phys. E: Sci. Instrum.* **22** 173–7

Huang S M, Stott A L, Green R G and Beck M S 1988c Electronic transducers for industrial measurement of low value capacitances *J. Phys. E: Sci. Instrum.* **21** 242–9

Huang S M, Xie C G, Thorn R, Snowden D and Beck M S 1992a Design of sensor electronics for electrical capacitance tomography *IEE Proc. G* **139** 83–8

Huang S M, Xie C G, Vasina J, Lenn C, Zhang B F and Beck M S 1992b Experimental evaluation of capacitance tomographic flow imaging systems using physical models *Proc. ECAPT'92* ed M S Beck, E Campograde, R C Waterfall and R A Williams (Southampton: Computational Mechanics Publications)

Husain A and Weisman. J 1987 Applicability of the homogenous flow model to two phase pressure drop in straight pipe and across area changes *AIChE Symp* pp 205–14

Irons G A and Chang J S 1983 Particle fraction and velocity measurement in gas-powder streams by capacitance transducer *Int. J. Multiphase Flow* **9** 289–97

Ishii M 1975 *Thermo-Fluid Dynamic Theory of Two-Phase Flow* (Paris: Eyrolles)

Ishii M and Grolmes M A 1975 Inception criteria for droplet entrainment in two-phase concurrent film flow *AIChE. J.* **21** 308–18

Jayet Y, Lakestan F and Perdrix M 1983 Simulation and experimental study of the influence of a front face layer on the response of ultrasound transducers *Ultrasonics* **21** 177–83

Kadambi V 1982 Stability of annular flow in horizontal tubes *Int. J. Multiphase Flow* **8** 311–28

Khan S H and Abdullah F 1991 Computer aided design of process tomography capacitance electrode systems for flow imaging *Sensors: Technology, Systems and Applications* ed K V T Grattan (Bristol: Adam Hilger) pp 233–8

—— 1992 Validation of finite element modelling of multi-electrode capacitive system for process tomography flow imaging *Tomographic Techniques for Process Design and Operations* (Southampton: Computational Mechanics Publications)

Kinghorn F C 1988 Challenging areas in flow measurement *Measurement and Control* **21** 229- 35

Kvernvold O, Vindoy V, Sontvedt T, Saasen A and Selmer-Olsen S 1984 Velocity distribution in horizontal slug flow *Int. J. Multiphase Flow* **10** 444–57

Lassahn G D, Stephens A G, Taylor D J and Wood D B 1979 *X-Ray and Gamma Ray Transmission Densitometry* (Research Report) (Idaho Falls, USA: Idaho National Engineering Laboratory, EG and G)

Lee S L and Durst F 1982 On the motion of particles in turbulent duct flows *Int. J. Multiphase Flow* **8** 125–46

Lemons R A and Quate C F 1979 Acoustic microscopy *Physical Acoustics* ed W P Mason (New York: Academic) pp 2–93

Liebmann G 1953 Electrical analogues *Brit. J. Appl. Phys.* **4**

Lin P Y and Hanratty T J 1986 Prediction of the initiation of slugs with linear stability theory *Int. J. Multiphase Flow* **12** 79–98

Lovinger A J 1985 Recent developments in the structure, properties and applications of ferroelectric polymers *Proc. 6th Meeting On Ferroelectricity, Japan J. Appl. Phys.* supp 24-2 (Kobe) pp 8–22

Lucas 1987 The measurement of two-phase flow parameters in vertical and deviated flow *PhD Thesis* University of Manchester

Lynnworth L C 1965 Ultrasonic impedance matching from solids to gases *IEEE Trans.*

SU **SU-12** 37–48

—— 1981 Ultrasonic flowmeters Pt.1. Eight types of ultrasonic flowmeters *Trans. Inst. Meas. Control* **3** 217–29

—— 1982 Ultrasonic flowmeters Part 2. Generation and propagation of pulses in single path countrapropagating flowmeters *Trans. Inst. Meas. Control* **4** 2–24

—— 1989 *Ultrasonic Measurements for Process Control* (Boston: Academic)

Mathur M P and Klinzing G E 1984 Flow measurement in pneumatic transport of pulverized coal *Powder Technol.* **40** 309–21

McQuillan K W and Whalley P B 1985 Flow patterns in vertical two-phase flow *Int. J. Multiphase Flow* **11** 161–75

Medlock R S and Furness R A 1990 Mass flow measurement—a state of the art review *Measurement and Control* **23** 100–13

Miller R W 1983 *Flow Measurement Engineering Handbook* (New York: McGraw-Hill)

Mishima Y and Ishii M 1984 Flow regime transition criteria for two-phase flow in vertical tubes *Int. J. Heat Mass Transfer* **27** 723–36

Moshfeghi M 1986 Ultrasonic reflection-mode tomography using fan-shaped-beam insonification *IEEE Trans. Ultrasonics, Ferroelectrics and Frequency Control* **UFFC-3** 299- 314

Naser-el-Din H, Masliyah J H and Nandakamur K 1990 Continuous gravity separation of concentrated bidisperse suspensions in an inclined plate settler *Int. J. Multiphase Flow* **16** 909- 19

Nigmatulin R I 1978 Multiphase Dispersed Flow (In Russian) (Moscow: Nauka)

Norton S J and Linzer M 1979 Ultrasonic reflectivity tomography: reconstruction with circular transducer arrays *Ultrasonic Imaging* **1** 154–84

Ohigashi H 1985 Piezoelectric polymers—materials and manufacture *J. Appl. Phys.* **24** 23–7

Ohigashi H, Koga K, Suzuki M, Nakanishi T, Kimura K and Hashimoto N 1984 Piezoelectric and ferroelectric properties of P(VDF-TrFE) copolymers and their applications to ultrasound transducers *Ferroelectrics* **60** 263–76

Oliemans R V A, Pots B F M and Trompe N 1986 Modelling of annular dispersed two-phase flow in vertical pipes *Int. J. Multiphase Flow* **12** 711–32

Omotosho O J, Frith B, Plaskowski A and Beck M S 1989 A sensing system for non-destructive imaging using externally compton-scattered gamma photons *Sensors and Actuators* **18** 1–15

Otake T, Tone S, Nakoa K and Mitsuhashi Y 1977 Coalescence and breakup in liquids *Chem. Eng. Sci.* **32** 377–87

Pitt G D, Entance P, Neat R C, Batchelor D W, Jones R E, Barnett J A and Pratt R H 1985 Optical fibre sensors *Proc. IEE* **132** 214–48

Pitteway M L V 1967 Algorithm for drawing ellipses or hyperbolae with a digital plotter *Computer J.* **10**

—— 1985 Algorithms of conic generation *Fundamental Algorithms for Computer Graphics* ed R A Earnshaw NATO ASI Series (Berlin: Springer) pp 219–37

Plaskowski A, Beck M S and Krawaczynski J S 1987 Flow imaging for multi-component flow measurement *Trans. Inst. Meas. Control* **9** 108–12

Pratt W K 1978 *Digital Image Processing* (New York: Wiley)

Pursley W C and Paton R 1985 The effect of air-oil mixtures on the performance of small turbine meters *Proc. Int. Conf. On Metering of Petroleum and Its Products* (London: Oyez Scientific and Technical Services Ltd)

Radon J 1917 Ueber die bestimmung von funktionen durch ihre integralwerte längs gewisser manningfaltigkeiten *Ber. Saechs. Akad. Wiss* **69** 262–78

Ruder Z and Hanratty T J 1990 A definiition of gas-liquid flow in horizontal pipes *Int. J. Multiphase Flow* **16** 232–42

Saeed 1987 Two component flow measurement using digital processing of optical signals *M. Phil. Thesis* Bolton Institute of Higher Education

Salcudean M, Chun J H and Groneveld D C 1983 Effect of flow obstruction on void distribution in horizontal air-water flow *Int. J. Multiphase Flow* **9** 91–6

Salkeld J 1991 Process tomography for the measurement and analysis of two-phase oil based flows *PhD Thesis* University of Manchester

Schafer M E and Lewin P A 1984 The influence of front-end hardware on digital ultrasonic imaging *IEEE Trans. on SU* **SU-31** 259–306

Schueler C F, Lee H and Wade. G 1984 Fundamentals of digital ultrasonic imaging *IEEE Trans. on SU* **SU-31** 195–217

Scott S L, Shoham O and Brill J P 1986 Prediction of slug length in horizontal large-diameter pipes *56 Annual Regional Meeting Papers* (AIME)

Seagar A D, Barber D C and Brown B H 1987 Theoretical limits to sensitivity and resolution in impedance imaging *Clin. Phys. Physiol. Meas.* **8** 13–31

Silk M G 1984 *Ultrasonic Transducers for Non-Destructive Testing* (Bristol: Adam Hilger)

Skelland A H P and Huang Y F 1977 Effects of surface active agents on fall velocities of drops *Can. J. Chem. Eng.* **55** 245–52

Soria A and De Lara H I 1992 Averaged topological equations for dispersed two-phase flows *Int. J. Multiphase Flow* **18** 942–63

Streeter V L 1961 *Handbook of Fluid Dynamics* (New York: McGraw-Hill)

Stuchley S S, Sabir M S and Hamid A 1977 Advances in monitoring of velocities and densities of particulates using microwave Doppler effect *IEEE Trans. Instrum. Meas.* **26** 21–4

Taitel Y, Barnea D and Dukler A E 1980 Modelling flow pattern transition for steady upward gas-liquid flow in vertical tubes *AIChE. J.* **26** 345–54

Taitel Y and Dukler A E 1976 A model for prediction flow regime transition in horizontal and near horizontal gas-liquid flow *AIChE. J.* **22** 47–55

Tarnoczy T 1965 Sound focusing lenses and waveguides *Ultrasonics* **3** 115–27

Tomita Y, Jotaki T and Hayashi H 1981 Wavelike motion of particulate slugs in a horizontal penumatic pipeline *Int. J. Multiphase Flow* **7** 151–66

Tsuji Y and Morikawa Y 1982 Plug flow of coarse particles in a horizontal pipe *ASME J. Fluid Eng* **104** 196–206

Von Ramm O T and Smith S W 1983 Beam steering with linear arrays *IEE Trans. on Biomedical Eng.* **BME-30** 438–52

Wang M, Dickin F J and Beck M S 1992 Improved electrical impedance tomography data collection system and measurement protocols *Tomographic Techniques for Process Design and Operations* ed M S Beck, E Campograde, R A Williams and R C Waterfall (Southampton: Computational Mechanics Publications)

Wang S K, Lee S J, Jones Jr O C and Lahey Jr R T 1987 3-D turbulence structure and phase distribution measurements in bubbly two-phase flow *Int. J. Multiphase Flow* **13** 327–44

Weigand F 1990 VLSI parallel processing in flow image based measurement *PhD Thesis* University of Leeds

Weigand F and Hoyle B S 1989 Simulations for parallel processing of ultrasound reflection- mode tomography with applications to two-phase flow meaasurement *IEEE Trans. Ultrasonics, Ferroelectrics and Frequency Control* **36** 652–60

Weszka J S 1978 A survey of threshold selection techniques *Computer Graphics, Vision and Image Processing* **7** 259–65

Willets I P, Azzopardi B J and Whalley P B 1987 The effect of gas and liquid properties on annular two-phase flow *Proc. 3rd Int. Conf. On Multiphase Flow* (The Hague: BHRA)

Woodhead S R, Barnes R N and Reed A R 1990 On-line mass flow measurement in pulverised coal injection systems *Powder Handling and Processing* **2** 123–7

Xie C G, Huang S M, Hoyle B S, Thorn R, Lenn C, Snowden D and Beck M S 1992 Electrical capacitance tomography for flow imaging: system model for development of image reconstruction algorithms and design of primary sensors *IEE Proc. G* **139** 89–98

Xie C G, Plaskowski A and Beck M S 1989a 8-electrode capacitance system for two-component flow identification. Part 1: Tomographic flow imaging *IEE Proc. A* **136** 173–83

—— 1989b 8-electrode capacitance system for two-component flow identification. Part 2: flow regime identification *IEE Proc. A* **136** 184–91

Yip F, Velar J and Gooier G 1979 The motion of small air bubbles in stagnant and flowing water *Can. J. Them. Eng.* **48** 229–35

Yorkey T J 1986 Comparing reconstruction methods for EIT *PhD Thesis* University of Wisconsin, Madison

—— 1990 EIT with piecewise polynomial conductivities *J. Comp. Phys.* **91**

Zabel T, Hansom C and Ingham J 1973 The influence of system purity, drop separation and heat transfer on the terminal velocity of falling drops in liquid-liquid systems *Trans. Inst. Chem. Eng.* **51** 162–4

Index

A-scan ultrasonic imaging 142, 144
Absorption attenuation coefficients 80
Accuracy 1, 4, 9, 19, 20
Acoustic imaging 111
Acoustic impedance matching 112, 115, 120
Acoustic lenses 111, 115, 116
Aggregative fluidization 40
Amplitude thresholding techniques 158, 170
Annular flow 32, 34, 37, 48, 49, 60, 63, 72, 79, 94, 131, 178, 179
Applied potential tomography (APT) 140
Area error 174, 178, 179
Argonne National Laboratory 15
Artificial neural network (ANN) 196
Asymmetrical flow 34
Automatic image analysis 161, 197
Automatic temperature compensation 13
Axial velocity 25, 197

Backprojection reconstruction algorithm 133, 140, 145, 196
Bandwidth 34, 89, 91, 92
Baseline drift in capacitance sensors 83, 84, 91, 92
Bimodal histogram 152, 153
Blurring of images 140, 168, 170
Boundary value problem 139
Bubble

agglomeration 45, 51
break-up 45, 67
coalescence 32, 36, 43, 57, 67
flow 34, 36, 37, 43, 152, 157
Reynolds number 54
velocity 43, 45
Bubbling fluidization 40, 44
Bulk velocity 12

CAD 86, 93, 121
Capacitance based component concentration techniques 83
Capacitance based flow imaging system for non-conducting fluids 84, 105
control and interfacing 99, 108, 120
computer aided design of electrodes 87, 92
electrode fabrication 98, 99, 188
performance parameters 95, 172
practical limitations 86
sensor electronic design 118, 194
Capacitance electrode field pattern 92, 115, 122
Capacitance flow noise 20
Capacitance sensors for flow imaging 81, 84, 87
baseline and sensitivity drift 83, 84, 87
computer aided design 92
electronic design 87
selection of parameters 92

sensor transfer function 90
Capacitance to voltage transducer 83,
      85, 87
CEA 15
Chain code transformation 161–163
Charge transfer principle 87
Chemical composition 31
Churn flow 36, 51, 186
Clamp-on flowmeters 13
Clamp-on sensors 115
Co-current gas/liquid flow in a
      horizontal pipe 37
   annular flow regime 38
   dispersed bubble flow regime 37
   intermittent flow regime 37
   stratified flow regime 37
Co-current gas/solids flow in a
      horizontal pipe 38
   dispersed flow regime 38
   moving cluster flow regime 38
   slug flow regime 39
   stationary cluster flow regime 39
   stratified flow regime 39
Component concentration
      measurement techniques 8, 18
   capacitance 8, 19
   optical absorption 8, 21
   radioactive attenuation 8, 18
   ultrasonic absorption 8, 20
Coal/oil slurries 15
Common mode rejection ratio 109
Component fraction measurement
      error 175
Component mass flow rate 5–7
Component ratio 4, 7
Component velocity measurement
      techniques 9
   cross correlation 8, 16, 17
   injected tracer 8, 14, 15
   laser Doppler 8–10
   pulsed neutron activation (PNA) 8,
      15
   nuclear magnetic resonance (NMR)
      8, 14, 15

microwave Doppler 8, 10–12
   ultrasonic Doppler 8, 12, 13
Conductivity sensors for flow imaging
      28
Constant current generator 108, 139
Contact impedance 108
Convolution-filter mask 153, 154
Core flow 94, 178
Coriolis flowmeter 8
   basic principle 22
   performance with two-component
      flows 23
   practical considerations 23
Contour following 162, 163, 167
Critical air velocity 71
Cross correlation function 17, 26
Cross correlator 18
Cross-correlation flow velocity
      measurement 8, 16, 17
Cross-sectional density distribution
      122
Current injection electrodes 106
Curve fitting 162

Damping characteristics of
      piezoelectric ceramics 116
Data acquisition system (DAS) 106,
      108
   demodulation and filtering 110
   differential input amplifier 109
   digital control 110
   voltage controlled current source
      (VCCS) 109
Data capture rate 178
Demodulator 110
Dense phase pneumatic conveying 68
Density measurement 4, 17, 22
Density meters 22
   gravimetric 22
   vibrational 22
Density profile 26
Dielectric spectroscopy 195
Differential input amplifier 109
Dilation techniques 160

Direct mass flowmeter 8
Direct measurement 8
Direct problem in image
    reconstruction 124
Discretization errors 94
Dispersed bubble flow 37, 42, 43
Doppler flowmeters 10
    laser 10, 11
    microwave 10, 12
    ultrasound 12, 13
Doppler shift principle 9
Drop Reynolds number 54
Drops and bubbles 54
    coalescence 57
    shape regimes 55
    terminal velocity (settling velocity)
        55
Dune flow 181
Dynamic range 109
Dynamic response 4, 19, 28
Dynamic threshold level 173

Edge point linking 162
Edge segmentation 162
Effluent stack gas measurement 21
Electrical capacitance tomography
    (ECT) 172, 180, 183, 184, 192
Electrical impedance imaging (EIT)
    system for conducting fluids
    105
    data acquisition system (DAS) 106
    electrodes 106
    interfacing electrodes to the DAS
        108
Electrical impedance tomography
    (EIT) 105
Electrical resistance tomography
    (ERT) 187, 192
Electromagnetic induction
    tomography (EMT) 194
Electrostatic interference 83, 84
Elongated bubble flow 95
Elongated churn flow 186
Engine flame front imaging 192

Entrainment 61
Eötvos number 55
Erosion techniques 160
European Concerted Action on
    Process Tomography (ECAPT)
    191
Expansion algorithms 160, 161

Fan beam projection 77
Fan shaped beam pattern 114
Fast fluidization 40
Ferrite circulator 11
Finite element modelling (FE) 93, 94,
    139
Fiscal measurement 1
Flow field 24, 25, 67, 171
Flow image categories 31
    deviated flow 33
    horizontal pipe stratified flow 32,
        33
    inclined flow 33 (see also deviated
        flow)
    separated planes 33
    vertical pipe cross-sectional image
        32
    vertical pipe multi-sectional image
        32
Flow imaging subsystems 28
    image reconstruction and display
        subsystem 29
    sensor signal processing subsystem
        28
    sensor subsystem 28
Flow maps 40
Flow pattern (see flow regime)
Flow regime 24, 34, 36, 37, 39, 40
Flow regime identification 41, 43, 52,
    75, 184, 197
Flow regime transition 42
    annular/intermittent flow 49
    from dispersed bubble flow 37
    stratified/annular flow 37, 48
    stratified/non-stratified flow 46
Flowmeter sales 1, 18

Fluidized bed 39, 183
Fluidized bed flow regimes 39
  bubbling fluidization 44
  fast fluidization 40
  particulate fluidization 40
  slugging regime 40
  turbulent regime 40
Forward problem 124 (see also direct
    problem)
Fractal techniques 41
Freeman chain code 161
  determining object area 165
  determining object perimeter 163
Fringing field effects 94
Froude number 46, 57

Gamma radiation sensor 2, 4, 80, 122
Gamma rays 18, 77, 80, 191
Gamma source 76
Gas/condensate flow 82
Gas/liquid flow 31, 42
Gas/oil flow 72
Gas/solid flow 24, 31, 38
Gas/water/solids separation 32, 33
Grey level histogram 151, 153
Guard electrodes 85, 177

Helical concentration sensor 20
High pass filtering 155
Hot film sensors 171
Hot wire sensors 171

Ill conditioned matrices 141
Image display 148, 151
  grey levels 151
  pseudo-colour techniques 151
  resolution 148. 150, 151
Image enhancement using filtering
    techniques 152
  basic concepts 152
  comparison of filtering techniques
    156
  using high pass filtering 155
  using Laplacian masks 156

using low pass filtering 155
Image interpretation 148
Image plane 25
Image presentation 150 (see also
    image display)
Image processing 148
Image reconstruction 122
  basic principles 122
  direct (or forward) problem 124
  inverse problem 124
Image reconstruction algorithms 126
  knowledge based 126
  backprojection 133
  Newton–Raphson 140
Image reconstruction system (IRS)
    106
Image resolution 148, 150, 151
Imaging system applications
  flow regime identification 184
  imaging gas flow in fluidized beds
    183
  imaging gas/solids flows 180
  measuring component concentration
    profiles 187
Impedance mismatch 118
Impedance spectroscopy 110, 195
Incipient fluidization 40
Incremental arc tracing algorithm 146
Incremental line tracing algorithm
    145
Inferential mass flowmeter 8
Inferential measurement 8
Injected tracer flow velocity
    measurement 15
Input signal resolution 175, 176
Instantaneous velocity 8
Integration time 28
Integration time constant 89
Interfacial effects 74
Interfacing electrodes 100
Interference fringes 9, 10
Intermittent flow 37, 38, 41, 49
Inverse problem in image
    reconstruction 124

Inverse Radon transform 123
Inviscid stress 59
Inviscid flow pressure 59
Isokinetic sampling system 4

Kelvin–Helmholtz instability 46, 59
Knowledge based reconstruction
    algorithm 126

Laminar flow 21
Laplacian masks 156
Laser Doppler flow velocity
    measurement 8
  differential method 10, 11
  reference beam method 10
Lateral velocity distribution 71
Lateral void fraction 71, 72
Leibnitz rule 63, 65
Liquid hold-up 50
Liquid/liquid flow 8
Liquid/solid flow 8, 24
Low pass filtering 154

Mass flow measurement 22–29
Mass flow rate 5–7, 70
Mathematical modelling of two-phase
    flow 61
  annular flow model 72
  averaging process 63
  dispersed bubble model 70
  particle flow model 70
  three-dimensional model 66
  unsteady two-dimensionsal slug
    flow 67
Maximum likelihood principle 126
Measurement period 17, 19
Measurement time 19
Meter factor 17, 18
Microwave absorption 2
Microwave Doppler flow velocity
    measurement 8, 10, 12
  bistatic configuration 10–12
  monostatic configuration 10, 11
Microwave horn antenna 10

Microwave tomography 194
Minimum fluidization 40
Missing data technique 126
Mixture velocity 7, 45, 51
Modulation 10, 20, 80
Moving cluster flow 38
Multibeam densitometer 19
Multimodality imaging systems 195
Multiple segment receivers 145, 146
Multiplexer 85

Newton–Raphson reconstruction
    algorithm 140
Non-destructive testing (NDT) 111,
    143
Non-homogeneous mixtures 24
Non-intrusive sensors 4
Non-steady three-dimensional flow
    66–68
Nuclear magnetic resonance (NMR)
    flow velocity measurement 8,
    14, 15, 191

Object detection 161
Oil/gas separator 34
Oil/water flow 126
Oil/water/gas mixtures 4
On-line calibration 10, 13
On-line water content measurement
    20
Open channel flow 32
Optical absorption component
    concentration measurement 8,
    21
Optical sensors for flow imaging 80
Orifice plate flowmeter 1, 9
Oscillator 92, 108, 110

Parallel projection 77
Particle absorption 20
Particle scattering 20
Particulate fluidization 40, 44
PE2D 94

Performance parameters in ECT systems 172
  input signal resolution 175, 176
  permittivity error 174, 175
  permittivity resolution 176
  signal to noise ratio 175
  spatial image error 174
  spatial signal resolution 176
Permittivity error 174, 175
Permittivity resolution 176
PET 191
Phase-sensitive demodulator (PSD) 110
Phased array sensors 111
Photomultiplier 10, 11
Piezoelectric ceramics 116
Piezoelectric polymer films 115
Piezoelectric transducer 12
Pixel cross-correlation 86
Planar ultrasonic sensors 115
Plug flow 181, 183
Pneumatic conveyor 180
Poisson's equation 139
Polarization effects 106
Polymer/copolymer films 115
Powder/gas flows 83
Process control using tomographic image data 197
Process tomography 31–33
Production platform 2
Programmable gain amplifier (PGA) 102, 109, 110
Projection 76, 103
Projection arc 143
Projection ellipse 144–146
Pseudo-colour techniques for image display 151
  density slicing 151
  grey level to colour level transformation 151
Pseudo-homogeneous flow 5, 6, 29
  mass flow rate 5
  volumetric flow rate 5

Pulsed neutron activation (PNA) flow velocity measurement 8, 15

Radial screens 93, 94, 97, 98
Radioactive attenuation component concentration measurement 8, 18
Radioactive particles 15
Radioisotope 16
Radon transform 123, 124
Ray sum 123
Rayleigh–Taylor instability 60
Real time measurement 2, 28, 32
Reflection mode ultrasonic flow imaging 114
Resistivity distribution 108, 141
Resolution 12 *et seq.*
Reynolds' rules 65
Reynolds number 51, 54
Roping 84
Rotating field concentration sensor 20, 21

Sampling cell 22
Sampling time 34
Scintillation detectors 15, 16
Search and shrink routine 131, 136
Sensitivity 28
Sensor classification 80
  hard field sensors 80
  soft field sensors 81
Sensors for flow imaging systems 82
  electrical capacitance 81, 83–85
  electrical conductance 81
  ionising radiation 80
  optical 80
  ultrasound 111, 114, 115
Separated flow 32, 33
Separation tanks 32
Separator 2, 33, 67
Series expansion methods 139
Settling time 87
Shell and tube heat exchanger 32, 40
Shrinking algorithm 160

Signal to noise ratio 175
Silicon technology in tomographic
    systems 195
Single-component flowmeter 2, 9
Single-component liquid flow 5, 6,
    24, 47
Slew rate 89
Sliding beds 7
Slug flow 36, 37
Slurry flows (see liquid/solid flows)
    9, 24
Solids loading 70
Solids mass flow rate 70, 180
Solids velocity 7
Spatial accuracy 176
Spatial averaging 20, 86
Spatial image error 174
Spatial resolution 12, 81, 82, 126, 140
Spatial signal resolution 176
Spray flow 41
Stability 125
Standing capacitance (see static
    capacitance) 86, 95, 97, 102
Static beds 7
Static capacitance 19
Stationary cluster flow 39
Stratified flow 32–34, 37, 41, 46, 47,
    51, 114, 132
Stray capacitance 83, 85
Subregions
    in intermittent flow 50
    in stratified flow 51
Subsea production system 2
Superficial gas velocity 71, 181
Superficial velocity 38, 40
Surcharging 33
Surface tension 7, 45, 60
Symmetrical field sensitivity 93

Taylor bubble 36, 43
Thickness mode 119
Three-component flow measurement 2
Three-dimensional image
    reconstruction algorithms 141

Three-dimensional modelling 66
Three-dimensional steady flow 67
Thresholding selection technique 158
Thresholding technique 158
    application to flow imaging 158
    selection 158
Time delay 17, 26, 28, 141
Time lag 17
Tomographic image analysis 197
Topological equations 66
Transducer (see sensor) 12 *et seq.*
Transducer transfer function 90
Transition boundary 45, 46, 180
Transition mechanisms 42, 50
Transit time 17
Transmission mode ultrasonic flow
    imaging 119, 141, 145
Transputer 99–103, 105
Transputer link adapter 99
Turbine flowmeter 2, 9
Turbulence 67
Turbulent flow 54
Twodimensional modelling 69
Two-component (multi-component)
    flow measurement 5, 7
    the basic problem 5
    conventional methods of
        measurement 7
    the importance of 2
Two-component flow imaging 24
Two-dimensional unsteady flow 67
Two-phase flow phenomena 34
    macroscale structure 36
    microscale structure 54
Two-phase fluid dynamics 31

U tube manometer 181
UKAEA 15
Ultrasonic absorption component
    concentration measurement 8,
    20
Ultrasonic Doppler flow velocity
    measurement 12
Ultrasonic flow imaging system 141

    choice of active element 116
    effect of wave propagation 118
    electronic system considerations
        120
    mechanical structure of the sensor
        117
Ultrasonic sensors for flow imaging
        111
    interaction with two-component
        flow 112
    limitations 111, 113
Upward gas/liquid flow in a vertical
        pipe 36
    annular flow regime 37
    bubble flow regime 36
    churn flow regime 36

Vector velocity imaging 197
Velocity distribution 24, 30

Velocity profile 26, 7
Venturi flowmeter 9
Viscosity ratio 55
Viscous shear 59
Viscous stress 59
Void fraction 7, 34, 51, 65, 67, 71
Void fraction measurement 7, 36, 67
Void fraction meter 19
Voidage mapping 67, 159, 192
Voltage to current converter 108, 109
Volumetric flow rate 5–7, 29

Wall effects 56, 57
Water content 4, 5, 83, 194
Water jet washer 22
Wave propagation effects 28, 118
Waves on the gas/liquid interface 59

X-band 11
X-rays 18, 77, 80, 171, 191, 196